LIST OF TITLES

Already published

A Biochemical Approach to Nutrition	R. A. Freedland, S. Briggs
Biochemical Genetics (second edition)	R. A. Woods
Biological Energy Conservation (second edition)	C. Jones
Biomechanics	R. McN. Alexander
Brain Biochemistry (second edition)	H. S. Bachelard
Cellular Degradative Processes	R. T. Dean
Cellular Development	D. R. Garrod
Cellular Recognition	M. F. Greaves
Control of Enzyme Activity	P. Cohen
Cytogenetics of Man and other Animals	A. McDermott
Differentiation of Cells	M. Bownes
Enzyme Kinetics	P. C. Engel
Functions of Biological Membranes	M. Davies
Genetic Engineering: Cloning DNA	D. Glover
Hormone Action	A. Malkinson
Human Evolution	B. A. Wood
Human Genetics	J. H. Edwards
Immunochemistry	M. W. Steward
Insect Biochemistry	H. H. Rees
Isoenzymes	C. C. Rider, C. B. Taylor
Metabolic Regulation	R. Denton, C. I. Pogson
Metals in Biochemistry	P. M. Harrison, R. Hoare
Molecular Virology	T. H. Pennington, D. A. Ritchie
Motility of Living Cells	P. Cappuccinelli
Plant Cytogenetics	D. M. Moore
Polysaccharide Shapes	D. A. Rees
Population Genetics	L. M. Cook
Protein Biosynthesis	A. E. Smith
RNA Biosynthesis	R. H. Burdon
The Selectivity of Drugs	A. Albert
Transport Phenomena in Plants	D. A. Baker

In preparation

Bacterial Taxonomy	D. Jones, M. Goodfellow
Biochemical Systematics	J. B. Harborne
The Cell Cycle	S. Shall
Gene Structure and Function	M. Szekely
Invertebrate Nervous Systems	G. Lunt
Membrane Assembly	J. Haslam

Editor's Foreword

The student of biological science in his final years as an undergraduate and his first years as a graduate is expected to gain some familiarity with current research at the frontiers of his discipline. New research work is published in a perplexing diversity of publications and is inevitably concerned with the minutiae of the subject. The sheer number of research journals and papers also causes confusion and difficulties of assimilation. Review articles usually presuppose a background knowledge of the field and are inevitably rather restricted in scope. There is thus a need for short but authoritative introductions to those areas of modern biological research which are either not dealt with in standard introductory textbooks or are not dealt with in sufficient detail to enable the student to go on from them to read scholarly reviews with profit. This series of books is designed to satisfy this need. The authors have been asked to produce a brief outline of their subject assuming that their readers will have read and remembered much of a standard introductory textbook on biology. This outline then sets out to provide by building on this basis, the conceptual framework within which modern research work is progressing and aims to give the reader an indication of the problems, both conceptual and practical, which must be overcome if progress is to be maintained. We hope that students will go on to read the more detailed reviews and articles to which reference is made with a greater insight and understanding of how they fit into the overall scheme of modern research effort and may thus be helped to choose where to make their own contribution to this effort. These books are guidebooks, not textbooks. Modern research pays scant regard for the academic divisions into which biological teaching and introductory textbooks must, to a certain extent, be divided. We have thus concentrated in this series on providing guides to those areas which fall between, or which involve, several different academic disciplines. It is here that the gap between the textbook and the research paper is widest and where the need for guidance is greatest. In so doing we hope to have extended or supplemented but not supplanted main texts, and to have given students assistance in seeing how modern biological research is progressing, while at the same time providing a foundation for self help in the achievement of successful examination results

General Editors:

W. J. Brammar, Professor of Biochemistry,
University of Leicester, UK

M. Edidin, Professor of Biology,
Johns Hopkins University, Baltimore, USA

Brain Biochemistry

H. S. Bachelard

Professor of Biochemistry,
St. Thomas's Hospital Medical School,
London

Second edition

Chapman and Hall
London and New York

First published in 1974
Reprinted in 1976
Second edition published in 1981 by
Chapman and Hall Ltd
11 New Fetter Lane, London EC4P 4EE
Published in the USA by
Chapman and Hall
in association with Methuen, Inc.
733 Third Avenue, New York, NY 10017

MyCopy version of the original edition 1981

British Library Cataloguing in Publication Data

Bachelard, Herman Stanton
Brain biochemistry. – 2nd ed. – (Outline studies in biology).
1. Brain chemistry
I. Title II. Series
599 .01′88 QP376

DOI 10.1007/978-94-009-5941-5
www.springer.com/mycopy

Contents

1 Introduction

The brain is the most complex and highly specialised of all mammalian organs. Understanding the complexity of its function remains man's greatest challenge. The functional unit is the neurone, or excitable nerve cell, making anatomical and chemical connections with other units in the system. Many of the essential biochemical connections of the nerve cell are dependent upon special morphological features: synaptic contact is mediated by chemical molecules, 'neuro-transmitters' which ensure the continued propagation of electrical impulses through sequential units of the system. Also closely related to the morphology of the nervous system is the chemical energy expended in maintaining distribution gradients of cations across cellular membranes. Chemical neurotransmission results in an alteration in cation distribution and while the energy-utilising mechanisms which underly their redistribution are not peculiar to the nervous system, they are of particular importance to neural function. The mechanisms of chemical transmission, in contrast, are peculiar to the nervous system.

Nerve cells are unique in their ability to trigger off and maintain conduction of electrical impulses over long distances, which may be measured in metres, without significant loss of strength of the conducted impulse. Remarkable also is the specificity of their connections, not only with other nerve cells, but also with non-neural target cells in sites such as the endocrine glands or muscles.

These unique features rest in the possession of semi-permeable excitable membranes which can be caused, rapidly and transiently, to undergo changes in permeability to small chemical molecules and to cations. The highly specialised nature of the constituent cells, with their unique function and specificity, is closely related to the structure of the whole tissue. The underlying chemical processes cannot be discussed or seen in perspective without constant awareness of related aspects of physiology and morphology. The brain is structurally extraordinarily complex in its distinct anatomical regions, each of which is heterogeneous in the types and structures of the constituent cells.

One aspect of the biochemical function of the brain can be seen in its efficient production of the energy required to support the unique processes referred to above. This energy, essentially stored as ATP, is produced from the oxidation of glucose by mechanisms common to all biological cells. The importance in the brain of these processes is quantitative, rather than qualitative. The brain depends absolutely for its ability to function normally on a constant supply of glucose and oxygen from the blood stream. It has virtually no reserves of chemical energy, compared with other tissues and organs. Stored concentrations of glucose and glycogen (each of the order of 1–2 μmoles/g) and of ATP (3 μmoles/g) are sufficient to maintain function in isolation for minutes only, if permanent damage is not to ensue and under normal circumstances, the brain cannot utilise alternative sources for its energy requirements [1]. The importance of the constant blood supply of essential nutrients can be readily appreciated if we remember that this organ, only some 3% of the total adult body weight, consumes some 20% of the glucose required by the whole body. This supply is in fact supported by the blood: one-fifth of the output of the heart passes through the brain. The brain is therefore the most sensitive part of the body to failure in oxygen or glucose. In the absence of either of these, fainting occurs within seconds, and if not corrected, coma and death follow rapidly. It is usually the first organ to suffer. Its peculiar sensitivity to abnormalities in energy metabolism can also be seen in the features of vitamin deficiency, especially of those vitamins such as the B group which function as coenzymes in intermediary energy metabolism. Although any deficiency affects the same metabolic pathways in the same way throughout the body, one of the most profound consequences is impaired mental function and in children, often mental retardation. It must be stressed that this is due, not to specialised qualitative metabolism by the brain, but to its very high sensitivity to any impairment in the normal processes of energy production.

This is of particular importance in the nutrition of the underdeveloped 'third world', where deficiency or dietary imbalance may cause irreparable mental damage to the developing child, and which has been the concern of special symposia [2, 3]. Not only is an inadequate environment increasingly suspected of leading to impaired intelligence in the poorer parts of the world, but evidence is also to hand that this can be seen in countries normally regarded as rich and developed. Although current discussions on the relative influences of heredity and of environment on the development of intelligence are heated and controversial, studies such as those on Scottish children over a 15 year period indicate that a consistent if small increase in intelligence can result from progressive improvement in their environment [4]. Further indications of the sensitivity of the brain to general metabolic impairment arise from the high proportion of inherited metabolic disorders which result in mental disturbance or retardation so important in Neurology and Psychiatry [5];

1.1 Regional cerebral metabolism

The great dependence of the brain on its supplies of glucose and oxygen, noted above, was originally assumed with good reason to be associated solely with energy production, yet early pioneering studies on the rates at which the normal adult brain uses these nutrients showed no difference with function. That is, the brain seemed to require the same energy whether the subject was responding to sensory stimuli, was thinking intensively or resting [6]. These studies were based on arterio-venous differences over the whole brain, and regional variations could not be assessed. Since then, measurements of regional variations in rates of cerebral blood flow have shown changes in specific regions in response to sensory stimuli or mental effort [7]. However this does not tell us if metabolic rates are changing in a similar manner with function: in epilepsy or induced fits in animals, overall rates of cerebral consumption of glucose and oxygen increase to a considerably greater extent than does the blood flow rate.

Some evidence for regional rates of glucose utilization in experimental animals is emerging from use of the autoradiographic 2-deoxyglucose technique [8], based on knowledge of the transport and metabolism of deoxyglucose in the brain [9]. While this provides much information about regional variations in metabolism with function in animals, it is subject to limitations in the types of function that can be studied. Furthermore it cannot be applied directly to studies in man, requiring as it does removal of the brain for autoradiography. The method has now been adapted for use in man. The ^{18}F-isotope emits positrons which penetrate the skull, so its localization within the brain can be monitored externally. The attachment of ^{18}F to the No. 2 position of the deoxyglucose molecule is thought not to affect its metabolism and it is being used, by positron emission computed tomography, to study regional cerebral glucose metabolism in conscious normal man [10]. It too has one essential limitation in that the isotope has a very short half-life, and a cyclotron is required to produce it in the vicinity of the research or clinical laboratory. Its expense renders it unlikely to come into general use, but so much promise in research and diagnosis is offered that one can foresee extensive use in selected specialist centres.

1.2 Cerebral requirements for glucose and oxygen

A direct relation between function and the requirements of glucose and oxygen for energy production is also being questioned in view of recent observations in man and experimental animals that mild hypoglycaemia and hypoxia show changes in the EEG and in behaviour without any perceptible energy deficit. This poses two possibilities: 1) that small regional variations in energy production are escaping detection; 2) that the brain is responding to changed availability of these nutrients in a protective homeostatic manner. The first possibility is regarded as an unlikely answer in itself (though it may contribute) because of the extraordinary rapidity with which an experimental animal can be aroused from deep hypoglycaemic coma by giving glucose: the metabolic machinery is thought not to be able to respond so fast. The second possibility is receiving much attention. This puzzling apparent anomaly is seen in a general sense if various aspects of energy metabolism and synaptic function are compared. We know that maintenance of synaptic function depends to a great extent on the cation pumping processes which require energy (Chapter 2).

Yet in addition to the conditions of hypoglycaemia and hypoxia noted above, treatment with various centrally active drugs (anaesthetics, depressants and excitants) may also cause changes in the EEG and in behaviour, and in synaptic function, without detectable change in the energy state [11, 12]. This has now been confirmed *in vitro*: electrical activity evoked from slices of hippocampus incubated *in vitro* can be affected markedly by slightly lower concentrations of glucose (2m*M*) which seem to have no effect on the intermediary metabolism of the tissue [13]. No clear pointers on the basis for this have emerged so far, but it may have important clinical significance in our understanding of such conditions as coma, stroke, epilepsy and also perhaps, dementia.

To be able to understand what he is trying to achieve, the biochemist who studies brain function must acquaint himself with related aspects of morphology, physiology and pharmacology; the chemical function of the brain cannot be separated from the architectural integrity of the cellular relationships. For a small book of this type the topics selected concentrate on the chemical events related to excitability and transmission and to the adaptability of the brain to react to various stimuli both from within and without the body. To this end, a brief description of the associated morphology and physiology seems an essential requirement and is treated first. Then follows a description of membrane permeability phenomena and neurotransmission. The final section of the book is concerned with aspects of the chemical response of the brain to its immediate environment and as a result of some of the hormonal signals reaching the brain.

References

[1] McIlwain, H. and Bachelard, H. S. (1971), *Biochemistry and the Central Nervous System* (4th ed.), Churchill, London.

[2] CIBA Foundation Symposium (1972), *Lipids, Malnutrition and the Developing Brain*, Elsevier, Amsterdam.

[3] DiBenedetta, C., Balażs, R., Gombos, G. and Porcellati, G. (eds) (1980), *Multidisciplinary Approach to Brain Development*, Elsevier, Amsterdam.

[4] Scottish Council for Research in Education (1949), *The Trend of Scottish Intelligence*, London University Press, London.

[5] Davison, A. N. (ed.) (1976), *Biochemistry and Neurological Disease*, Blackwells, London.

[6] Kety, S. S., Woodford, R. B., Harmel, M. H., Freyhan, F. A., Appel, K. E. and Schmidt, C. F. (1948), 'Cerebral blood flow and metabolism in schizophrenia: effects of barbiturate semi-narcosis, insulin-coma and electroshock', *Am. J. Psychiat.*, **104**, 765–770.

[7] Lassen, N. A. and Ingvar, D. H. (1972), 'Radio-isotopic assessment of regional cerebral blood flow', *Progress in Nuclear Medicine* **1**, 376–409.

[8] Sokoloff, L., Reivich, M., Kennedy, C., De Rosiers, M. H., Patlak, C. S., Pettigrew, K. D., Sakurada, O. and Shinohara, M. (1977), 'The (^{14}C)-deoxyglucose method for the measurement of local cerebral glucose utilization: theory, procedure, and normal values in the conscious and anesthetized albino rat', *J. Neurochem.*, **28**, 897–916.

[9] Horton, R. W., Meldrum, B. S. and Bachelard, H. S. (1973) 'Enzymic and cerebral metabolic effects of 2-deoxy-D-glucose', *J. Neurochem.*, **21**, 507–520.
[10] Reivich, M., Kuhl, D. and Wolf, A. (1977), 'Measurement of local cerebral glucose metabolism in man with ^{18}F-2-fluoro-2-deoxy-D-glucose', *Acta Neurol. Scand.*, **56** (Suppl. **64**), 188–190.
[11] Bachelard, H. S. (1981), 'Cerebral Metabolism and Hypoglycaemia', In: *Hypoglycaemia* (ed. Marks, V. and Rose, F.) Blackwells, Oxford.
[12] Siesjö, B. K. (1978), *Brain Energy Metabolism*, Wiley, Chichester.
[13] Cox, D. W. G. and Bachelard, H. S. (1980), 'The effect of lowered glucose on field potentials evoked from the hippocampal slice *in vitro*', *Neuroscience Letts.*, Suppl. **5**, S448.

2 Appearance of the brain

2.1 Gross appearance

Biochemists tend to study the brains of small mammals and consciously or subconsciously extrapolate to what might occur in the human brain, itself subject to obvious limitations in opportunities for chemical exploration. Yet what do we mean by the 'mammalian' brain, since the brain of a rat or guinea pig is obviously far different in appearance and many functions from that of Man? The brain has evolved and specialised within mammalian species more than any other organ of the body: Fig. 2.1 shows a comparison of the brains of a selected group of mammals and it should be remembered that increasing size is not necessarily associated with increased intelligence or sophistication of function. The main discernible change that has occurred during evolution of the mammalian brain is in the size and complexity of the cerebral cortex. The increase in surface area per unit of volume of the cortex has been effected by increased folding so that the convolutions of the human cerebral cortex are considerably more extensive than of the rat or rabbit. The function of the cortex has altered also: the 'primary cortex' concerned with sensorimotor function (Fig. 2.1) has remained proportionately much the same, but the areas devoted to 'association', i.e. areas concerned with higher functions of learning and decision-taking, have increased considerably [1].

Other areas, such as the limbic system (Fig. 2.1), concerned with more primitive functions of homeostasis, motivation and especially emotion [2], are phylogenetically older, and have changed little in relative size, as the shaded areas show [3, 4]. For the non-anatomically trained, the nomenclature of the regions and specialised parts of an organ as complex as the brain is daunting. In fact the brain should

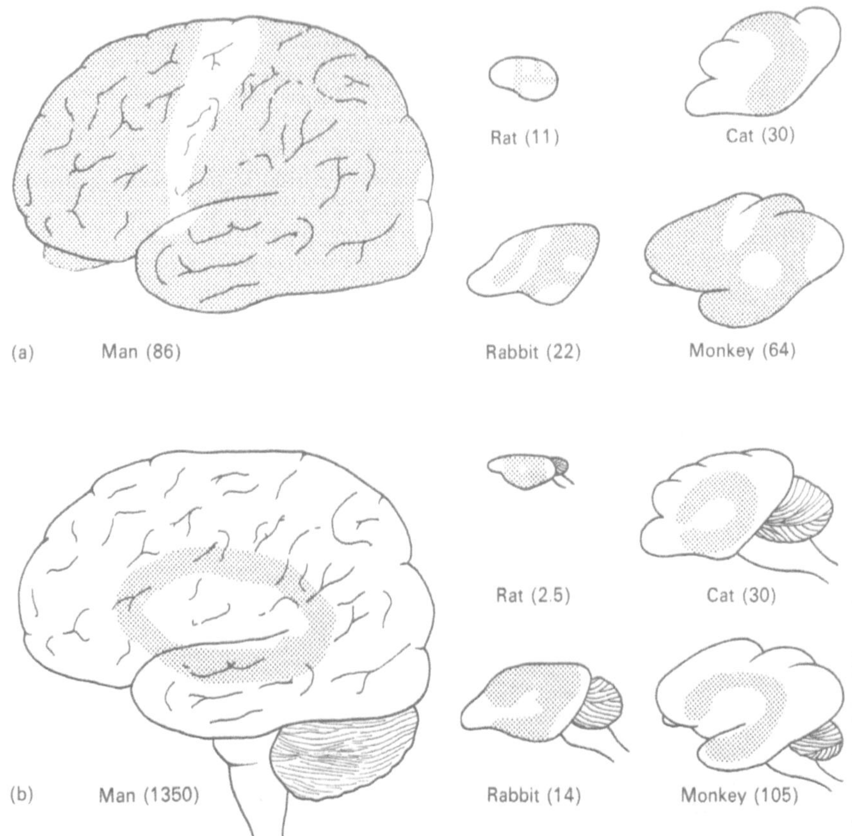

Fig. 2.1 Comparison of some functional areas of the brains of various mammals. **(a)** Proportions of sensorimotor and association (shaded) areas. Values in parenthesis are percentages of association cortex. **(b)** Schematic representation of the relative areas of the limbic system. Values in parenthesis are the weights of the brains in grams. The drawings are approximately half-size (linear).

really be regarded as a collection of highly specialised organs rather than as a single organ. Often various areas of the brain are cited in biochemical articles as experimental material and the untutored reader may be uncertain of the site and significance of the part named. The major areas of the human brain which are likely to be referred to are shown in Fig. 2.2 (see also Table 3.2 of Chapter 3). The first part of the illustration (a) shows the lateral aspect of the whole brain, viewed from the left hand side: the next drawing (b) shows some of the internal parts which are seen if the brain is divided into the two hemispheres. This is the right hemisphere viewed from the left hand side. The whole brain has been divided into four main parts for convenience: the cerebrum, the cerebellum, the mid-brain and the brain stem, and it is the latter which contains a large number of specialised parts indicated in Fig. 2.2. The third illustration (c) shows the interior of the brain cut horizontally and viewed from above.

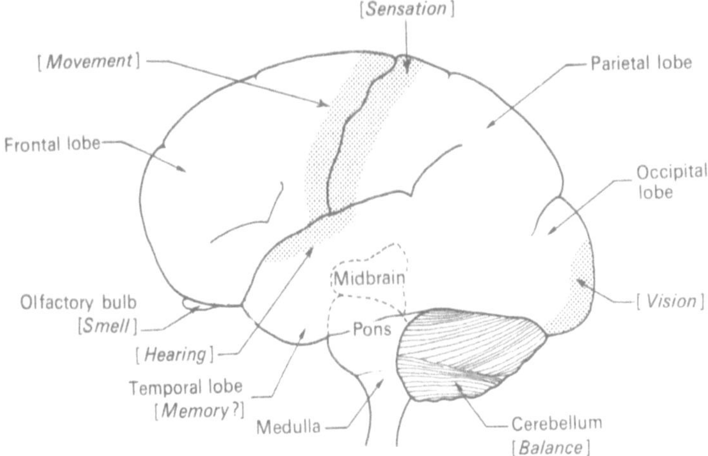

Fig. 2.2 Drawings of the human brain. **(a)** View from the left side, showing major areas with some indication of function.

2.2 Fluid compartments

One prominent feature of the gross anatomy of the brain is its extensive blood supply–perhaps not surprisingly, since it uses about one-fifth of the total blood used in the body. The blood volume of the brain is only about 3% of the total brain volume and the efficiency of the supply is ensured by the extensive ramified system of capillaries. Exchange of solutes between the various fluid compartments (blood, cerebrospinal fluid and the extracellular tissue space) and the cells themselves, exhibits features not always found in other parts of the body. The most studied of these features is the 'blood-brain barrier'. Originally this concept arose from the limited penetration of injected dyestuffs from the bloodstream to the brain substance, which was also found to occur with a variety of small highly water-soluble chemicals of a wide range of classes: sugars such as fructose and sucrose, and charged molecules such as thiocyanate and most amino acids. Microscopic examination of the endothelial cells of the walls of the blood capillaries of the brain indicated that they were packed more tightly together than in capillaries outside the brain, so there seemed to be a sound basis for a physical permeability barrier at the capillary wall. This suffices for relatively large molecules like proteins but is inadequate to explain the apparent limited permeability of substances such as glutamate. Indeed if radioactive glutamate is present in the blood-stream it equilibrates rapidly with the glutamate within the brain, to judge from the extent of its labelling, but a massive increase in concentration of the external glutamate does not materially change its internal concentration. For substances like glutamate, and many others (including precursors of amines with specific neural function, Chapter 3), the 'blood-brain barrier' can be regarded as a homeostatic mechanism whereby the internal concentration is maintained by active processes of extrusion [5].

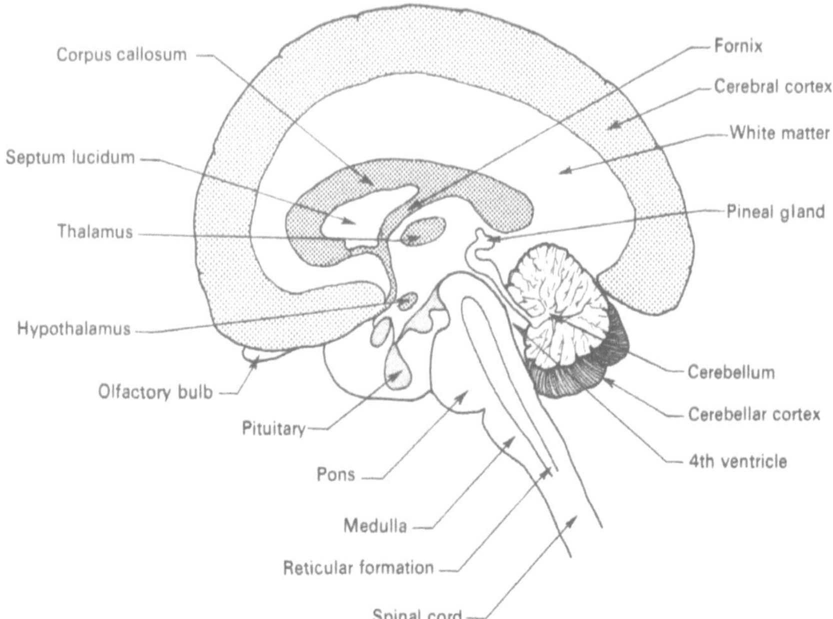

Fig. 2.2 Drawings of the human brain. **(b)** Cut-away view as in **(a)** showing some of the internal features.

The peculiarities of the 'blood-brain barrier', which is believed to fulfil a protective role for the highly sensitive brain, can therefore impede potential treatment of brain disorders by rendering difficult the internal accumulation of drugs or metabolites. Such a difficulty might be overcome by a knowledge of the biochemistry of the system at fault. Thus in Parkinson's disease, a degenerative and progressive disorder associated with muscular tremor and akinesia, anatomical observation showed degeneration of certain nerve tracts and histochemical analysis revealed a parallel loss of dopamine (see Chapter 3). Dopamine (3, 4-dihydroxyphenylethylamine) is one of those substances which are not easily transported into the brain, but its immediate metabolic precursor, Dopa (dihydroxyphenylalanine), is readily taken up. Marked improvement in many patients has been achieved by treatment with Dopa.

2.3 Microscopic appearance

The first appreciation of the morphology and cytology of nervous tissues, the network of individual cells and their processes, came towards the end of the nineteenth century with the application of the improved light microscope and the development of new staining methods. One of these, based on silver salts, causes selective staining of neuronal cell bodies and their processes in thin sections with such clarity that the structures seem to stand out in almost a three-dimensional picture. This is the Golgi stain. An alternative method, the

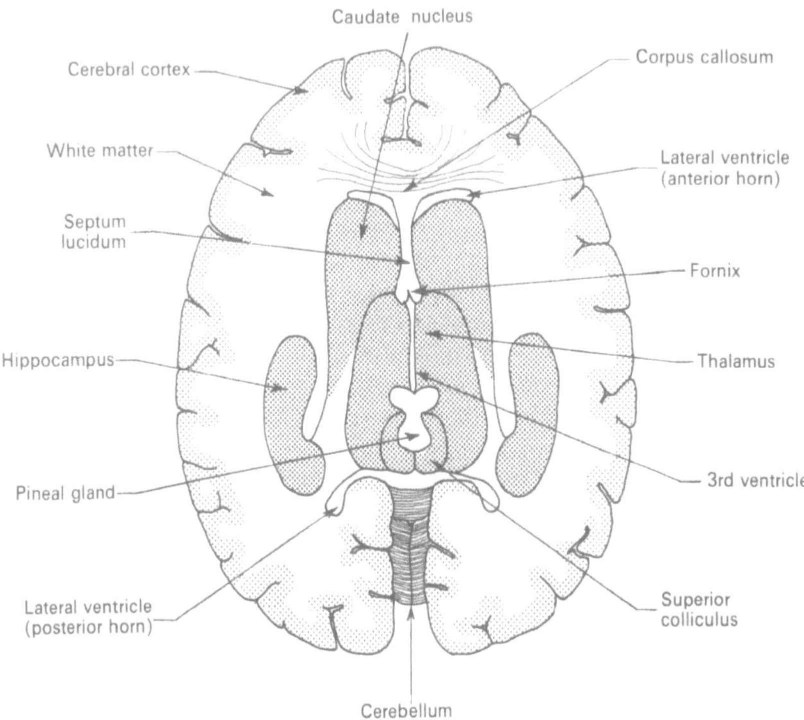

Fig. 2.2 Drawings of the human brain. **(c)** Cut-away view from above, with the front of the brain to the top of the drawing.

Nissl stain, shows the cell bodies of neurones and glia but not their processes. The observations, made using the light microscope, of cell bodies, axons, dendrites and dendritic spines (see below), were sufficient to lead to the 'neuronal hypothesis' with the concept of synaptic junctions some seventy-five years ago [6]. Further insight into fine structure, especially of the synapses, and confirmation of the neuronal hypothesis, had to wait until the advent of the electron microscope some forty years ago.

Light microscopy had clearly demonstrated the occurrence of a variety of cell types, classified into two main groups, neurones (the excitable nerve cells) and glial cells (non-excitable). Within each group, different types have been discerned.

2.3.1 Neurones

These (Fig. 2.3) may have large or small cell bodies (*perikarya*) but all are characterised in possessing a large *nucleus* containing a prominent *nucleolus*, a high content of *ribosomes* in the cytoplasm (either free or attached to an extensive *endoplasmic reticulum*) and a high content of *mitochondria*. Such features are compatible with active synthetic and secretory activities and the large capacity for energy production referred to in Chapter 1. Essential characteristics are the prominent processes which form extensions of the outer cell membrane:

axons and dendrites. Axons are usually long, relatively thin, and emerge from a swelling in the cell body – the axon hillock. The axons are sometimes branched and usually, but not always, covered by an insulating sheath, the *myelin* sheath, consisting of a spiral (giving the impression of concentric rings in cross section) of membranes. Myelinated axons form the main routes for the efficient rapid conduction of the electric impulse from the neurone (efferent) to another part of the system and the connections are made through synapses (below). Dendrites are usually thicker, shorter and highly branched, do not have a myelin sheath and carry the impulse from synapses to the nerve cell (afferent). These processes contain *neurotubules*, apparently identical with the microtubules of the mitotic apparatus and of contractile tissues, and are thought to be associated with axonal transport of materials from the perikaryon through the axon (Chapter 3).

Three main types of nerve cell can be identified by means of their processes. '*Unipolar*' cells contain only one axon and examples of these are sensory cells of ganglia. '*Bipolar*' cells have two processes, an axon and a dendrite, and are found as sensory receptor cells concerned with sight, smell, and hearing. The majority of the neurones are *multipolar*, having one axon and many dendrites. Multipolar cells fall into two main classes, named according to their shapes: the pyramidal cells of Fig. 2.4 and stellate cells.

2.3.2 Glial cells

Glial cells (Fig. 2.5) do not possess the excitable characteristics of the nerve cell, are generally smaller, but also have processes emanating from their cell bodies. These processes are relatively short and often highly branched. There are three main types. *Astrocytes* often occur close to blood vessels: their processes terminate in 'end-feet' which make contact with the blood capillary wall. These are thought to be concerned with nutrition, possibly acting as mediators in the transport of materials from the blood stream to the neurones. Indeed a highly specific means of causing degeneration of glial cells without direct and immediate damage to the neurones, is by promoting hyperammonaemia. This can occur naturally, as a result of severe liver damage, or experimentally by the portocaval shunt technique [7]; the astrocytes become swollen and vacuolated. The *oligodendroglia* are also satellite cells and are intimately concerned in the central nervous system with the myelin sheath of the axon, which they produce. The third group, the *Schwann* cells, perform the same function in myelination of peripheral nerves outside the brain. Fig. 2.6 is a diagrammatic representation of the process of myelination. The myelinating cell wraps itself around the axon so that its plasma membrane forms a spiral. The nucleus of the cell can be seen lying close to the axon. Each of the myelinating cells (oligodendroglia or Schwann cells) forms a unit of myelin along part of the length of the nerve and many such may be required for the

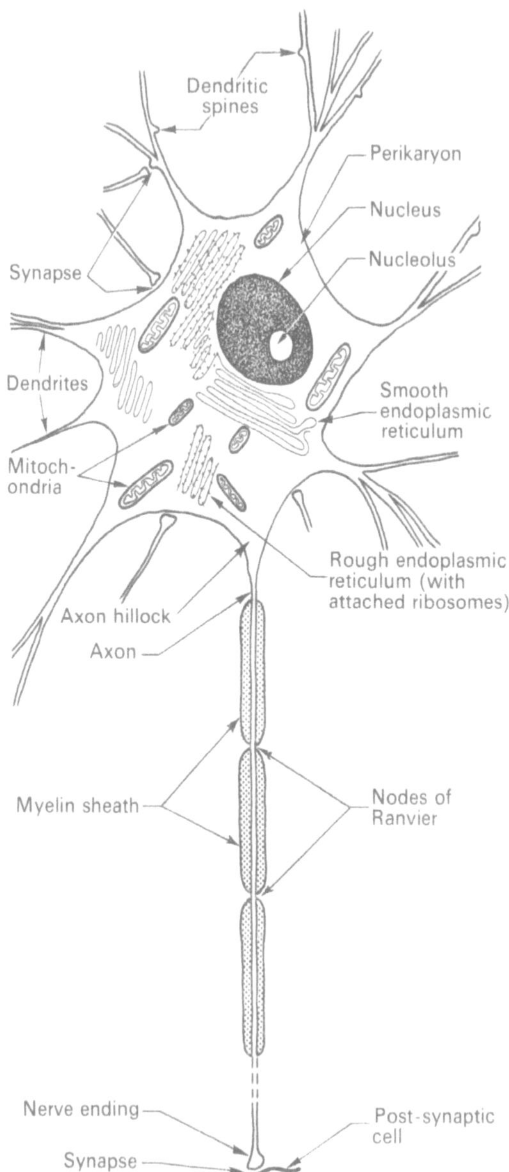

Fig. 2.3 Schematic drawing of a neurone.

entire length of that nerve. At the points where the myelin from one glial cell ends and that from the next begins, is a small gap where the nerve is not covered by the sheath: the 'node of Ranvier' (Fig. 2.3, see also Chapter 3.)

2.3.3 The synapse

The junction of one nerve cell with another (or of a nerve with innervated target cells such as in muscle or the endocrine glands) has been

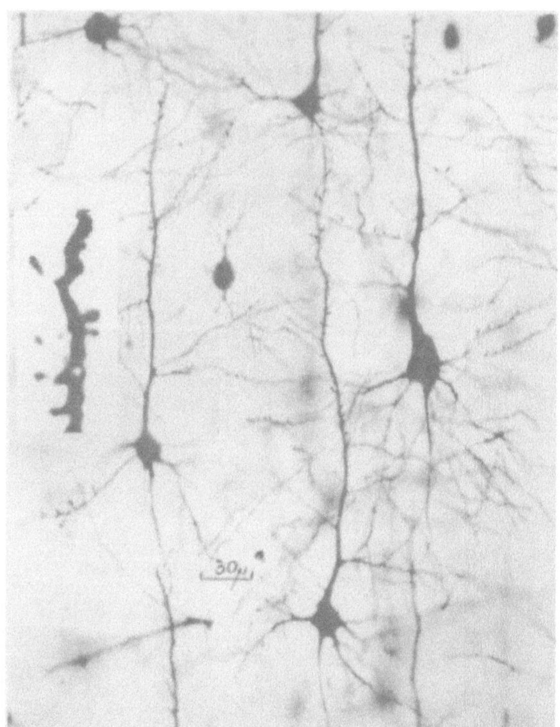

Fig. 2.4 Light microscopic picture of neurones. The section of the cerebral cortex was stained by the Golgi method, magnification x360. The inset shows a higher magnification (x2870) of the dendritic spines. The photograph was kindly supplied by Professor E. G. Gray, University College, London.

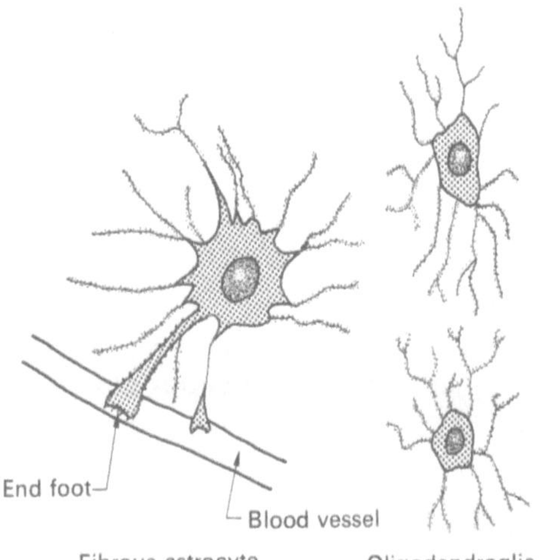

Fig. 2.5 Glial cells.

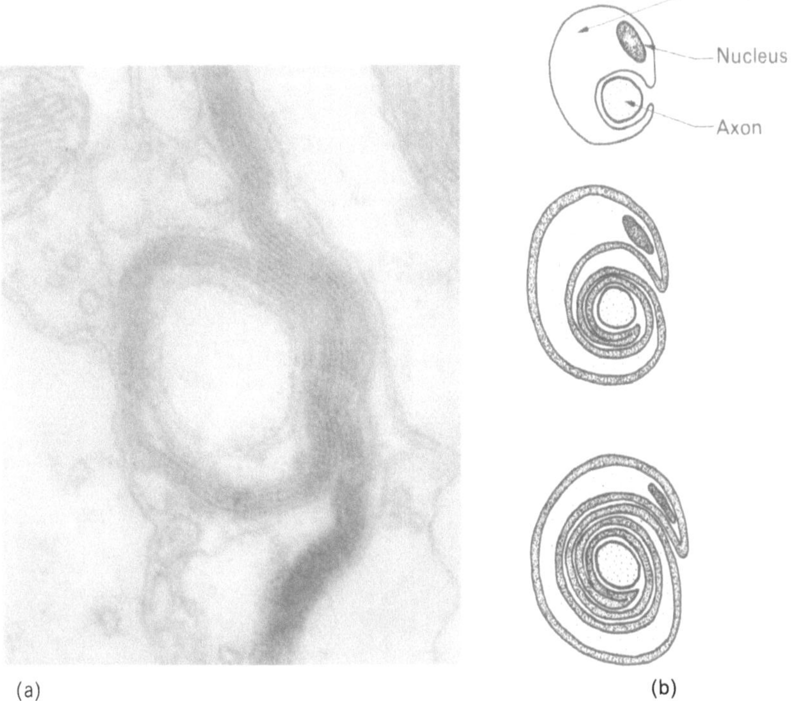

Fig. 2.6 Formation of the myelin sheath. **(a)** An electron micrograph of rat cerebral cortex showing myelinated axons, containing mitochondria. Magnification x 35000. **(b)** During the process of myelination, the glial cell (a Schwann cell in the peripheral nervous system or an oligodendroglial cell in the central nervous system) wraps itself around the axon and with a spiral motion, forms the concentric circles of myelin from its outer plasma membrane.

noted to be at the specialised *synapse* (Fig. 2.7). As the axon approaches its point of contact with the subsequent or *post-synaptic* cell, it enlarges into a specialised structure, known as the *nerve ending*. It is completely surrounded by membrane which, with few exceptions, is not fused with the membrane of the postsynaptic system, but is separated from it by a gap, some 200Å in width, known as the *synaptic cleft*. It is across this gap that chemical mediation of nerve transmission occurs (Chapter 3). The exceptions, where the pre-synaptic and post-synaptic membranes are fused, are known as *electrical synapses*. These have been positively identified only rarely [8]; so far they seem to occur in motor synapses in primitive systems such as the earthworm and crayfish, with spinal electromotor neurones of the electric fish and with ciliary ganglionic neurones in the chick. For the major part, especially in mammalian systems, synaptic transmission is chemically-mediated and the synapses are of the generalised structure shown in Fig. 2.7. Their internal morphology is characterised by pre-synaptic mitochondria and *synaptic vesicles*, believed to store the chemical transmitter molecules, and by the post-synaptic apparatus

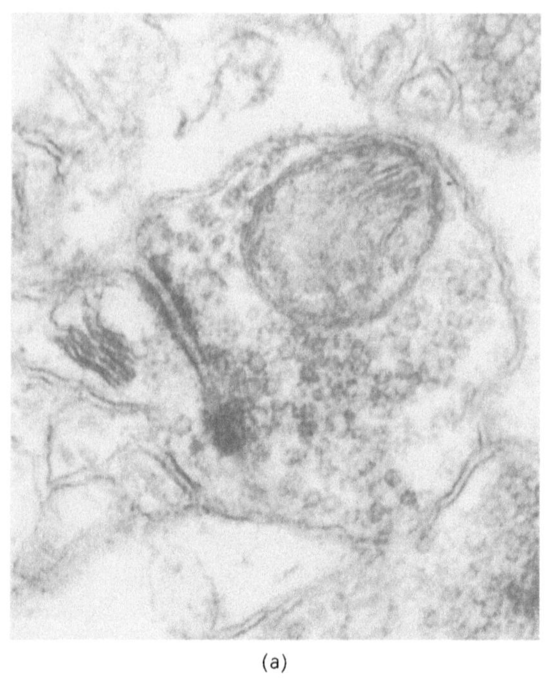

(a)

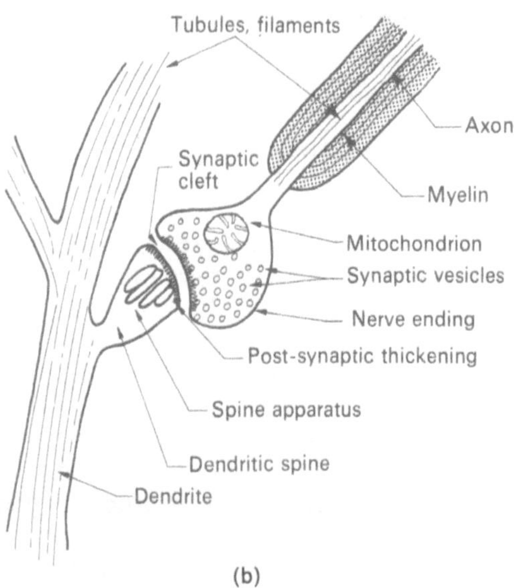

(b)

Fig. 2.7 The synapse. **(a)** Electron micrograph of a dendritic synapse in rat cerebral cortex, magnification x 28 175 (Courtesy of Professor E. G. Gray, University College, London). **(b)** Schematic drawing of a dendritic synapse, illustrating the constituent parts.

depicted. The synaptic vesicles may be smooth as shown in Fig. 2.7 and are usually about 500Å in diameter. Granular vesicles are also known to occur and may be of similar size or larger, up to 1000Å. Synaptic vesicles of different appearance are believed to be associated with adrenergic transmission, involving catecholamines, rather than cholinergic transmission, involving acetylcholine (Chapter 3). The post-synaptic system may be the cell body of another neurone, or the dendrite of another neurone, where the dendritic post-synaptic membrane is often swollen to form the spine of Fig. 2.7 (see also Fig. 2.3).

The molecular events surrounding the process of chemical transmission have stimulated much interest for biochemists, and are now described.

References

[1] Campbell, H. J. (1965), *Correlative physiology of the nervous system*, Academic Press, London and New York.

[2] White, L. E. (1965), A morphological concept of the limbic lobe. *Int. rev. Neurobiol.*, **8**, 1–34.

[3] Cajal, S. R. (1955), *Studies on the cerebral cortex: limbic structures.* (transl. by Kraft, L. M.), Lloyd-Luke, London.

[4] McLean, P. D. (1954), Studies on limbic system (visceral brain) and their bearing on psychosomatic problems, in *Recent Developments in Psychosomatic Medicine* (ed. Wittkower, E. D. and Cleghorn, R. A.). Pitman, London, pp. 101–125.

[5] Bradbury, M. (1979), *The Concept of a Blood-Brain Barrier*, Wiley, Chichester.

[6] Sherrington, C. (1906), *The integrative action of the nervous system.* Cambridge University Press, Cambridge.

[7] Cavanagh, J. B., Lewis, P. D., Blakemore, W. F. and Kyu, M. H. (1972). Changes in the cerebellar cortex in rats after portocaval anastomosis. *J. neurol. Sci.*, **15**, 13–26.

[8] Phillis, J. W. (1970). *Pharmacology of Synapses*. Pergamon, London.

3 Neurotransmission

3.1 The resting potential

To understand the principles of neurotransmission, we need some grasp of the bioelectric properties of the excitable nerve cell membrane. All cells are surrounded by semi-permeable membranes with a disequilibrium of the charged molecules on each side of that membrane, to give a charged field. Thus all cells have a difference in electrical potential across their outer cell membranes, in the range of −10 to

−90 millivolts, resulting from the relative distribution of ions between the intracellular and extracellular regions, and are said to be 'polarised'. The ions concerned are mainly K^+ and negatively charged macromolecules inside the cell, with Na^+ outside. The resting potential (−60 to −70 millivolts in most neurones) has been concluded to be due to the peculiar permeability properties of cell membranes: they are considerably more permeable to K^+ and Cl^- than to Na^+. This selective permeability to these ions may be a matter of size since the hydrated sodium ion is some 50% larger than the hydrated potassium ion. Nerve cells, in common with most cells, contain high concentrations of K^+ (100–120m*M*) and low concentrations of Na^+ (20m*M*) relative to the concentrations existing outside in the extracellular fluids (5m*M*K^+ and 140m*M*Na^+). The cell membrane therefore separates two compartments with unequal concentrations of NaCl and KCl in each. From a knowledge of these concentrations, the potential difference across the membrane can be calculated.

If the concentrations of Na^+, K^+ and Cl^- inside and outside the cell are known, an approximation to the resting membrane potential is given by the *Goldman equation* [1]:

$$E = \frac{RT}{F} \ln \left[\frac{P_K K_0 + P_{Na} Na_0 + P_{Cl} Cl_i}{P_K K_i + P_{Na} Na_i + P_{Cl} Cl_0} \right]$$

where R = gas constant, T = absolute temperature and F = Faraday (the electric charge/g. equivalent of a monovalent ion). P_K, P_{Na} and P_{Cl} are the permeabilities of the ions and K_0, Na_0, Cl_0 and K_i, Na_i, Cl_i are their concentrations outside and inside respectively. Since Cl has been found to contribute only slightly to the membrane potential, the equation is simplified to:

$$E = \frac{RT}{F} \ln \left[\frac{P_K K_0 + P_{Na} Na_0}{P_K K_i + P_{Na} Na_i} \right]$$

$$\text{or } E = \frac{RT}{F} \ln \left[\frac{K_0 + b Na_0}{K_i + b Na_i} \right], \text{where } b = \frac{P_{Na}}{P_K}$$

Since R, T and F are constants, at 37° and converting to $\log_{10}$, the relation becomes

$$E(\text{millivolts}) = 62 \log_{10} \left[\frac{K_0 + b Na_0}{K_i + b Na_i} \right].$$

If the extracellular and intracellular cerebral concentrations of the ions (m*M*) are taken as $K_0$5, Na_0 140, K_i 112, Na_i 20, and b as 0.04 (2), E becomes − 60mv which is similar to the membrane potential measured directly for cerebral neurones.

The Goldman equation is derived from the *Nernst equation*, as the sum of the Nernst equations for each ion species. It gives the equilibrium potential due to the asymmetric distribution of each ion across the semipermeable membrane:

$$E = \frac{RT}{F} \ln \frac{A_0}{A_i}$$

where A_0 and A_i are the activity coefficients of the ion species outside and inside. For practical purposes, since activity coefficients are not known, concentrations are used instead. However the Goldman equation is an approximation only: it assumes passive diffusion of the ions through the membrane, or that active movement of one ion species is coupled to active movement of another, i.e. it assumes that movement of ions is 'electroneutral' rather than 'electrogenic'. The basis for the derivation of these equations is given by Hodgkin and Katz [la], see also Woodbury [1b] and a previous book in this series by Davies [3].

We know that active cation transport does occur and produces a 'steady state' situation in the resting cell, where ion concentrations and potentials are maintained, so that net ion fluxes (active plus passive) are zero. Efflux of Na^+ has been shown to take place against concentration gradients of Na^+ and requires external K^+. It is an energy-consuming process and movement of K^+ is coupled to movement of Na^+ in the opposite direction, not necessarily on a 1 : 1 basis. The ratio for transport across the erythrocyte membrane is approximately $2K^+$ transported for $3Na^+$ [4], and while this is likely to vary, a similar ratio may be the case elsewhere, including the brain. Certainly part of the influx of K^+ is linked to the efflux of Na^+ in the brain. The tendency for ions to diffuse across the membrane, K^+ out and Na^+ in, is countered by the active transport of the cations 'uphill', i.e. against concentration gradients so that Na^+ is pumped out into the extracellular environment of high Na^+; K^+ outside moves to the high K^+ environment within the cell. So any passive leakage of ions is compensated by the continued expenditure of energy of the active transport process. On the discharge of the excitable cell (Section 3.3) when more rapid efflux of Na^+ occurs, this cation redistribution then becomes an integral part of the recovery process and is effected through participation of a membrane-bound enzyme. This enzyme, which perhaps uses as much as one-third of the energy produced metabolically in the brain and stored as ATP [2], hydrolyses the ATP to produce ADP and inorganic phosphate, and is the metabolic basis for the 'sodium pump' (below).

Thus the development of the resting membrane potential results essentially from the efflux of Na^+ and from the differential permeability of the membrane to K^+ and Na^+. The necessary unequal distribution of these cations is maintained at the expenditure of energy by the sodium pump.

3.2 The sodium pump

Membrane-bound 'ATPases' were recognised for some years before Skou [5] suggested this activity to be associated with active transport of Na^+. He worked with crab nerve and his results were shortly followed by an immense study of the activity in many tissues; the brain was found to be particularly active [2]. Association of the Na^+, K^+-adenosine triphosphatase with active cation transport rests on convincing if circumstantial evidence.

(1) full enzymic activity requires both Na^+ (100m*M*) and K^+ (6m*M*) in concentrations comparable to the extracellular Na^+ and the extracellular K^+ (above). The Michaelis constants found for both cations for the enzyme are similar to those calculated for cation transport.

(2) the cation specificity of the enzyme is identical to that for monovalent cation transport (e.g. NH_4^+ can replace K^+ but not Na^+).

(3) high concentrations of Na^+ are inhibitory to K^+ in both processes.

(4) ouabain (strophanthin g) in similar micromolar concentrations inhibits both processes to a similar extent.

Perhaps the most convincing evidence arose from studies on reconstituted erythrocyte 'ghosts', in which the internal contents of the cell can be partially replaced. This elegant work demonstrated the 'vectorial' properties of the ATPase: the activity of the enzyme was increased by higher internal Na^+ [6]. The enzyme reaction involves intermediate phosphorylation of the enzyme (Fig. 3.1). After some speculation on its nature (the intermediate was originally suspected to be a phosphorylated serine residue), experiments using ATP labelled on the γ-position with ^{32}P, with hydroxylamine which removes the phosphate, and studies on the enzyme activity against synthetic substrates, together provided strong evidence that it is an acyl phosphate. Subsequently, use of tritiated propyl hydroxylamine enabled the group to be isolated and identified as a γ-glutamyl phosphate [7].

3.3 The action potential and nerve conduction

The unique feature of the excitable cell is seen when the 'resting state' is upset. The earliest technique, still used, was to apply an electrical impulse to the cell by means of an electrode. This pulse caused 'depolarisation'. If a red blood cell or a liver cell were to be stimulated in this way the depolarisation of the membrane is seen in the slow passive loss of the potential difference across it. If a nerve cell is stimulated, a very different series of events ensues. The potential changes from about −60 mV to −70 mV but this does not continue in the same way as it does in the non-excitable cell: there is a rapid overshoot to the extent that the potential may become positive, to

$$\text{Enz.} + \text{MgATP}^{2-} \xrightarrow{Na^+} \text{E:P} + \text{MgADP}^-$$

$$\text{E:P} \xrightarrow{K^+} \text{E} + \text{P}$$

$$[\text{Overall reaction:- } \quad \text{ATP} \longrightarrow \text{ADP} + \text{P}]$$

Fig. 3.1 Na^+, K^+-Adenosine Triphosphatase.

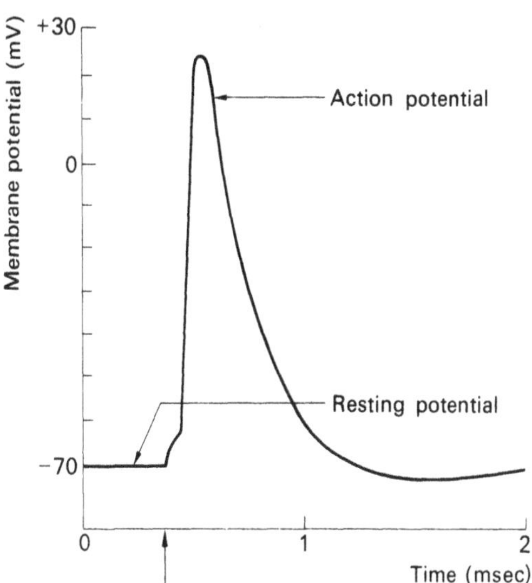

Fig. 3.2 An action potential. The arrow shows the time at which the stimulus was applied.

some + 10 to + 30mV (Fig. 3.2). This is associated with movement of cations: the membrane becomes more permeable to Na^+ [1]. Presumably some change has occurred in the porosity of the membrane so that permeability to Na^+ becomes less restricted. This could result from changes in the conformation of macromolecules constituting the lipo-protein matrix of the membrane. For convenience, although we do not understand the mechanisms, we think in terms of an increasc in the 'pore-size' of the membrane, i.e. the pores which under the resting state were too small to permit passage of Na^+, have now increased in size to the extent that rapid passage of Na^+ can occur. Sodium flows into the cells and potassium out until near electrochemical equilibrium is reached. This depolarisation usually lasts about half a millisecond and is known as the *action potential.*

The increases in permeability to Na^+ and K^+ are not simultaneous: Na^+ permeability increases first and initially to a greater extent. Subsequently Na^+ permeability decreases and K^+ permeability increases; the process becomes self-limiting and the potential difference returns to its original value. The system remains inexcitable for a few milliseconds, the 'refractory period', and this transient period is considered to result from a condition when changes in membrane permeability are such that it is temporarily non-permeable to Na^+ and freely permeable to K^+. This knowledge came from use of the 'voltage-clamp' technique which enabled the investigator to prevent the uncontrolled explosive occurrence of the action potential and to control changes in the membrane potential, during which current and ion flux rates could be measured [8]. The process can be summarised as follows: excitation causes first an increase in permeability to Na^+, which flows in. This is followed by a decrease in the permeability to

Na^+, coupled with increased permeability to K^+. The cell has the capacity for many repeated depolarisations before its cation balance reaches the stage where no further excitation can occur. This does not occur normally because energy is continually being expended to return the cations to their original distribution. This re-distribution is effected through participation of the Na^+, K^+-ATPase, described above.

The action potential has certain clear characteristics. There is a critical size of the stimulus which produces it, smaller stimuli having no effect and larger stimuli producing no greater effect. The critical size is the *threshold* and the process is *all-or-none*, in that the size of the action potential is independent of the size of the stimulus, provided it is at or above the threshold. However this does not mean that the 'signal' is uncontrolled. Control at the nerve cell which is depolarised may be exerted by the sequence of the stimuli in frequency and in size, i.e. the *latency*, the interval of time between stimulation and the action potential, may be decreased with repeated stimuli of increasing size and the frequency of the stimuli can effect the post-synaptic response (Section 3.6). Other factors also operate: the properties of the axon which conducts the signal, of the nerve endings and their chemical transmitter, and of the post-synaptic system. Once the nerve cell has 'fired' the action potential is conducted along the axon very rapidly (the rate may be as high as hundreds of metres per second) until it reaches the nerve ending where chemical transmission usually produces the post-synaptic response. Non-myelinated axons, which are not 'naked' but surrounded by a membrane of glial origin, are generally short and some loss or dissipation of the amplitude of the impulse may occur. Continued renewal of the propagated impulse is a feature of myelinated axons which may be many centimetres in length. The myelin sheath acts as a very efficient insulator and there is little loss of strength of signal in the axon between the nodes of Ranvier. It is at these nodes, the gaps in the myelin sheath, were the signal is renewed. Provided that the action potential reaching the node is still above threshold, the excitable axonal membrane is exposed to and surrounded by the extra-cellular fluid; depolarisation again occurs. The action potential moves through the next myelinated section. The action potential can therefore pass down the entire length of a myelinated axon without decreasing in size, by 'leaping' from node to node and being continually reinforced. The rapidity of the conduction depends very much on the diameter of the axon and on the thickness of the myelin sheath.

3.4 Chemical events at the synapse

It should be noted that we know quite well what happens, but we have very little understanding of the precise mechanisms which underly chemical neurotransmission. When the action potential reaches the nerve ending, it causes release of specialised chemicals almost certainly from their storage sites in the synaptic vesicles. This

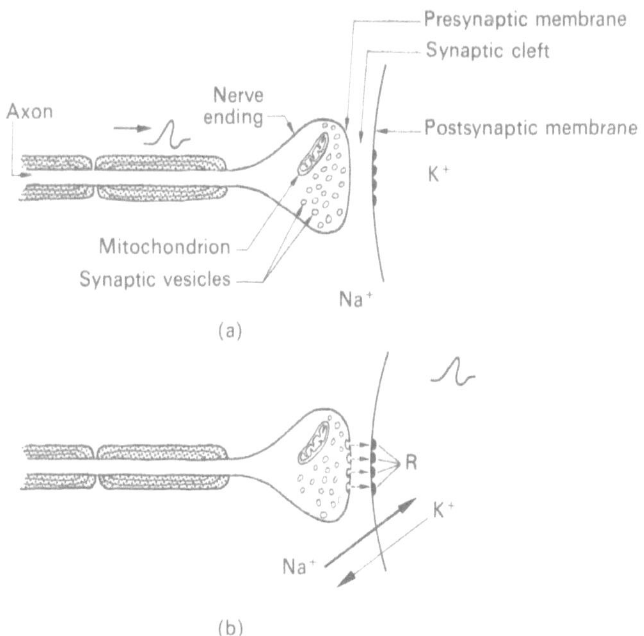

Fig. 3.3 Schematic drawing of synaptic transmission. On arrival of the action potential (∿) at the nerve ending **(a)**, transmitter molecules are released and react with receptors (R) on the post-synaptic membrane. The permeability of the membrane to Na^+ and K^+ changes **(b)**.

release can also be caused by local application of strong solutions of potassium salts in experiments with isolated preparations. There is some doubt as to the storage sites of the chemical transmitter molecules, and for acetylcholine there appear to be at least two, and possibly three, sites available: a 'free' site and two 'bound' sites. One of these, a tightly-bound site, does not seem to be the source of the active transmitter which is believed to come from the other, a relatively 'loosely-bound' site. Both bound sites are considered to reside on or in the synaptic vesicles. The free site may act as a reserve pool outside the synaptic vesicles in the cholinergic synaptosomal cytoplasm. Again there is argument and speculation about the way in which the transmitter is released from the synaptic vesicles into the synaptic cleft. Three theories have been advanced: (1) that the whole vesicle passes into the cleft where it disintegrates, discharging its contents ('exocytosis'), (2) that the vesicle, on coming into contact with the pre-synaptic membrane, opens up into a pore in the synaptic membrane through which the transmitter passes, (3) less widely held now, that the stimulus causes the synaptic vesicle to discharge its contents into the nerve-ending in the immediate vicinity of the pre-synaptic membrane and the transmitter diffuses through that membrane. There is some morphological and histochemical evidence in support of exocytosis of noradrenaline vesicles in adrenergic systems, especially from observations of release of stoichiometric amounts of ATP and protein with the catecholamine. Whatever the mechanism in cholinergic systems (most workers favour exocytosis), it is certain that the transmitter is released into the 200 Å wide synaptic cleft (Fig. 2.3 and 3.3) and diffuses across it to react with specific receptor sites on the post-synaptic membrane. The result is a change in ion permeability of the post-synaptic membrane to cause either the depolarisation or hyperpolarisation described below under 'post-synaptic events' (Section 3.6).

We do not know how the reaction of the transmitter with its post-synaptic receptors causes the response; all we can really do is to assume that, by mechanisms unknown, the reaction causes a change in membrane permeability to cations, perhaps by modifying the conformation of the lipoprotein matrix of the membrane, to open an 'ionophore' (Section 3.6). This is not unreasonable since there are many examples in biology of changes in protein conformation resulting from specific interactions with simple small molecules. Much of the recent effort of biochemists and biophysicists has been devoted to identification and isolation of these receptors (Section 3.6). It is an incredibly difficult task because the special function of the receptor could well reside in its architectural relationship with the membrane in which it is situated.

3.4.1 Identification and occurrence of neurotransmitters

The first and best characterised of the chemicals identified is acetylcholine (Table 3.1) mainly from elegant studies on the neuromuscular junction [9]. It fulfils the requirements initially regarded as the essential criteria for identification of a neurotransmitter, although it must be noted that these criteria were formulated largely from evidence which accumulated for cholinergic systems! These were:

(1) The chemical must be stored in the nerve endings from which it is released.

(2) It must be released upon pre-synaptic stimulation and shown to be present in the extra-cellular fluid in the vicinity.

(3) When applied post-synaptically it must mimic the action seen when the pre-synaptic system is stimulated.

(4) Specific antagonists should be recognised which prevent the action of both the chemical and electrical stimulation. This usually means a pharmacological agent which blocks the interaction of the transmitter with its receptor.

(5) Mechanisms for destruction of the transmitter in the post-synaptic region were thought to be essential to limit its duration of action (see below).

Historically the first indications of the role of acetylcholine came from observations that it mimicked the action of stimulation of parasympathetic nerves and a few years later that it was released as a result of stimulation of the vagus nerve. The classical experiments were on the neuromuscular junction, where Dale and his co-workers [10] demonstrated both its release on stimulation and that its applica-

Table 3.1 Likely transmitters in the mammalian nervous system.

Name	*Structure*	*Antagonist*	*Some sites*
Acetylcholine	$CH_3COOCH_2CH_2\overset{+}{N}(CH_3)_3$	Curare, atropine*	Neuromuscular junction Autonomic ganglia Caudate nucleus
Noradrenaline	HO, HO (ring) $-CHCH_2NH_2$ with OH	Various†	Lower brain stem Hypothalamus Substantia nigra
Dopamine	HO, HO (ring) $-CH_2CH_2NH_2$	Haloperidol, Spiroperidol	Corpus striatum Caudate nucleus
Serotonin	HO (indole ring, NH) $CH_2CH_2NH_2$	Methergoline	Hypothalamus Caudate nucleus
Glycine	$HOOCCH_2NH_2$	Strychnine	Spinal cord Cerebellum
γ-Aminobutyrate	$HOOCCH_2CH_2CH_2NH_2$	Bicuculline	Spinal cord Cerebellum Cerebral cortex

* Two receptor types have been recognised in cholinergic systems: 'nicotinic', blocked by curare and 'muscarinic', blocked by atropine.

† Two receptor types have also been recognised in adrenergic systems: 'α', blocked by many alkaloids (e.g. the ergot group and related compounds) and 'β', blocked by e.g. propranolol or tropolones. It is possible that these are not necessarily different sites: α and β receptors may be the same receptor site with properties which vary according to localised conditions.

tion mimicked the action of neurotransmission there. Demonstration of the blockage of transmission by antagonists such as curare soon followed. The importance of rapid subsequent destruction of acetylcholine was recognised from the effects of anticholinesterase agents in potentiating nerve stimulation and by the lethal effects of such inhibitors of the enzyme involved, *acetylcholinesterase*, which destroys the transmitter by hydrolysis. The remaining criterion (storage) awaited recognition of the nerve ending by electronmicroscopy and its separation as a distinct entity which was achieved in the early 1960's [11].

If the cells of most tissues are disrupted by carefully controlled homogenisation in aqueous iso-osmotic media, the internal cellular constituents (nuclei, mitochondria, soluble cytoplasm and fragments of membranous material from the outer cell membrane and the endoplasmic reticulum) can be separated reasonably well by centrifugation in centrifugal fields of increasing force. If similar techniques are applied to the brain, the 'mitochondrial' fraction which results consists of many fragments in addition to mitochondria, including myelin fragments and nerve ending particles. The latter, named 'synaptosomes', can be isolated due to a fortuitous circumstance: on gentle disruption of the tissue the axon breaks near the point where it swells to form the nerve ending and the broken membrane apparently re-seals to produce an ending which is usually intact. The majority of these isolated nerve endings do not seem to be 'leaky', at least as far as soluble enzymes such as lactate dehydrogenase are concerned [12]. Chemical and enzymic analysis of the synaptosomes revealed that they were enriched in acetylcholine and also in the enzyme responsible for its synthesis (Fig. 3.4). If the synaptosomes are disrupted by osmotic shock (by resuspension in water instead of the iso-osmotic or hyper-osmotic environment in which they have been prepared) the synaptic vesicles can be collected and shown to be rich in acetylcholine or other transmitters.

One of the most important of the above criteria to be satisfied has been thought to be the need for rapid destruction of the transmitter, due to the dire consequences if this is prevented in cholinergic systems. However other chemicals, now considered to be neurotransmitters, fulfil many of the criteria except this one – there is no obvious mechanism for their rapid destruction. These are amines; noradrenaline, dihydroxyphenylethylamine ('dopamine'), 5-hydroxytryptamine ('serotonin'), amino acids: e.g. glycine and γ-aminobutyric acid, or various peptides (Fig. 3.10). None of these has a true enzymic equivalent to the esterase which destroys acetylcholine. All of the degradative enzymes are slower acting than acetylcholine esterase and some of the amines and amino acids have been shown to be readily transported across membranes [13]. It is now felt that diffusion away from the synaptic cleft is sufficiently rapid to allow cessation of their activity, particularly since aminergic systems show generally slower responses than cholinergic systems. The release of all transmitters shows an

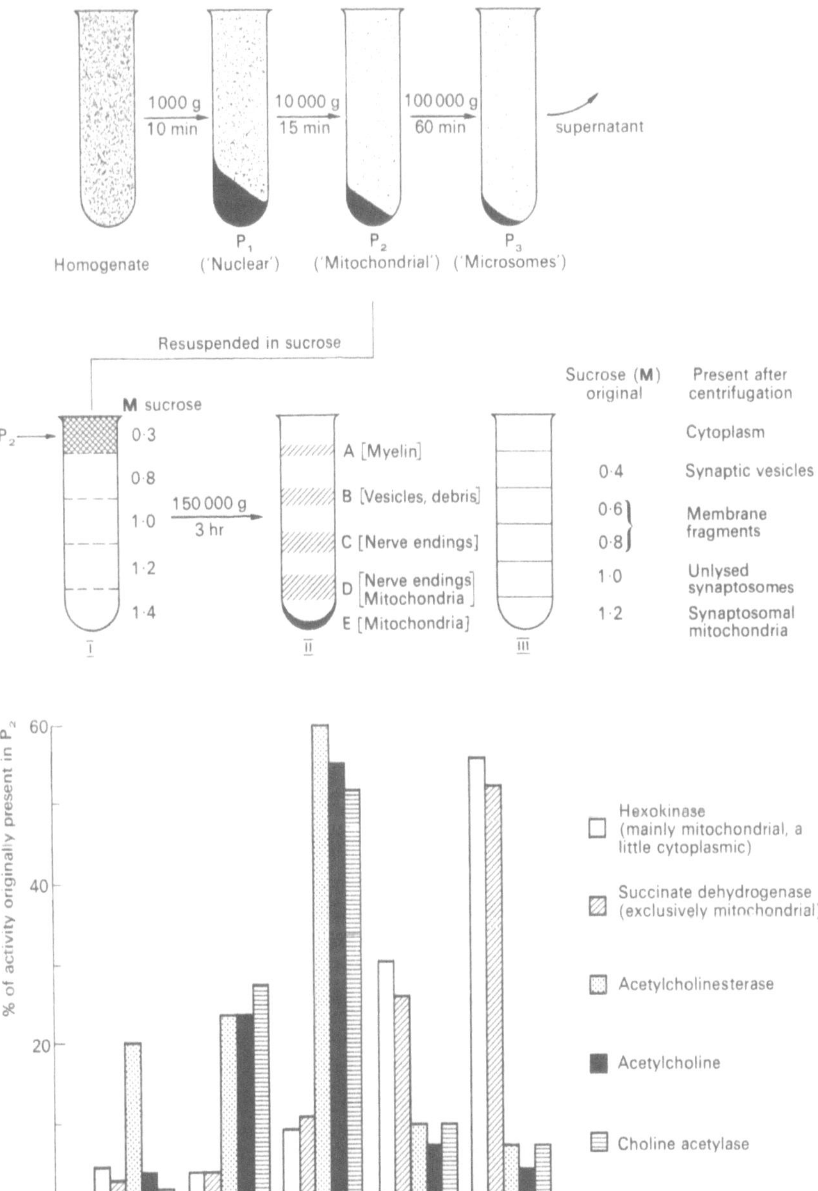

Fig. 3.4 Preparation of synaptosomes and synaptic vesicles [11]. The crude mitochondrial fraction (P_2), obtained by centrifugation of a sucrose homogenate, is resuspended in 0.3 M sucrose and layered over a stepwise gradient of sucrose (I). Centrifugation in swing-out buckets gives the separation shown in (II). Resuspension of subfraction C in water lyses the constituent synaptosomes to release the contents, which can be separated by density-gradient centrifugation (III). Chemical and enzymic analyses of the subfractions of (II) are shown in (IV).

absolute requirement for Ca^{2+} and is blocked by Mg^{2+} *in vitro*, causing paralysis in nerve-muscle preparations.

The main transmitters that have been identified are shown in Table 3.1 and Fig. 3.10. They vary very considerably in their distribution throughout the nervous systems; for example, acetylcholine is most concentrated in parts of the brain stem, the spinal cord and in the ganglia of the autonomic system (Table 3.2). This transmitter occurs in lower concentrations in the cerebellum and in minute amounts in the hippocampus, in contrast to the amines shown in Table 3.2. Whereas noradrenaline is most highly concentrated in the sympathetic system and the hypothalamus, dopamine occurs in the greatest amounts in the caudate nucleus, the corpus striatum and the basal ganglia of the brain stem. Serotonin (5-hydroxytryptamine) occurs mainly in the hypothalamus, the caudate nucleus and the pineal gland.

Knowledge of the identity of some of the transmitters in various areas and regions of the nervous system has emerged also from elegant histochemical studies using fluorescent microscopy. The methods are based primarily on exposure of thin sections of frozen or freeze-dried tissue to formaldehyde vapour [14]. Dopamine and noradrenaline are converted to dihydroisoquinolines (Fig. 3.5) which fluoresce with a green or yellow-green light of wavelength about 470 nm. Serotonin and other amines produce a yellow fluorescence with a higher wave-length (520 nm). By such techniques suspected amine transmitters can be specifically detected within the limits of resolution of light microscopy, i.e. with precision in different small regions of the brain, but with less precision within the cells.

3.4.2 The quantum hypothesis

The suggestion [15] that chemical transmitters are released in discrete packages or 'quanta' resulted from quantitative studies on the minute, random, spontaneous electrical discharges detected at the neuromuscular junction, known as *miniature endplate potentials.* These occur at rates of about 1 per second and have an amplitude of about 0.5 millivolt. The hypothesis has received much support and is believed to hold, not only for the neuromuscular junction, but also for chemical synapses everywhere, including those in the central nervous system. Miniature end plate potentials show the same qualitative properties as the action potential in cholinergic systems: they are blocked by curare (which blocks the receptor) and they are enhanced in both amplitude and duration by inhibitors of acetylcholinesterase (below) which normally would remove the transmitter by destroying it. The conclusion is that the miniature end plate potentials arise from reactions of small amounts of transmitter with post-synaptic receptors. They are well below threshold so no action potential ensues, or in the case of muscle, no contraction occurs. They are additive, and analysis of the distribution of amplitudes of the spontaneous potentials and those evoked by stimulation showed that they were integral multiples of a unit component, the 'quantum'. One quantum is the amount of

Table 3.2 Regional Distribution of Neurotransmitters

Region	*Part*	*Acetylcholine*	*Noradrenaline*	*Dopamine*	*Serotonin*
Cerebrum	Cortex	**15–25**	1·2 (0·3)	0·3 (0·5)	2 (0·2)
	Olfactory bulb	9	0·3		2·5
Cerebellum		1·5	0·6	0·1	0·5
Mid brain	Superior colliculi	**30**	1·2	**10**	
	Substantia nigra		(1)	(7)	(3)
Brain stem	Hippocampus	≈ 0	1	4.5	3
	Caudate nucleus		0·6 (0·6)	**50 (25)**	**9 (2)**
	Corpus striatum		0·1	**(25)**	0·2
	Basal ganglia	**50**	1	**45**	
	Lateral geniculate	**25**	0·6		
	Pineal gland				**50–100***
	Hypothalamus	15	**8 (6)**	3·5 (5)	8 (3)
	Thalamus	20	1 (1)	0·3 (1·5)	4 (1)
	Pons	20	1·2	0·7	2.5
	Medulla	15	2.5 (0·6)	0·7 (0·1)	3·5 (2)
Spinal Cord	Dorsal, ventral horns	15	1·5	≈ 0	3
	Ventral roots	**100**			
	Dorsal roots	1			
Peripheral nerve	Sciatic	35			
Ganglia	Superior Cervical	**200**	≈ 0		≈ 0

Values are approximate only, as nmoles/g fresh weight, from dog and cat, and those for human brain are given in parenthesis [2, 13, 24]. The richest sources are indicated by heavy type. There is considerable species variation: small rodents such as rat give higher values generally. Many of these regions are shown in Fig. 2.2.
* Values in the pineal gland are subject to diurnal variation: high by day, low at night (Chapter 4).

acetylcholine which produces one miniature end plate potential. Arrival of the nerve impulse at the nerve ending causes the release of much greater total amounts of transmitter and is due to a massive increase in the number of quanta, rather than an increase in the amount of transmitter released per quantum.

The number of molecules of transmitter per quantum has proved difficult to assess. By preventing replenishment of acetylcholine (using hemicholinium, a reagent which prevents choline uptake), repeating stimulation until all of the acetycholine is depleted, and knowing the total amount of transmitter available or stored pre-synaptically, the amount of acetylcholine used per quantum can be roughly calculated from the number of quanta required to produce the action potential at a neuromuscular junction. About 200 to 300 quanta are released per impulse at mammalian neuromuscular junctions, which amount represents less than 0.5% of the total available transmitter. Each quantum was calculated to contain about 50 000 molecules of acetylcholine [15]. The possibility that each quantum of acetycholine represents the content of one synaptic vesicle seems attractive, though unlikely: independent measurements [16] of the acetylcholine con-

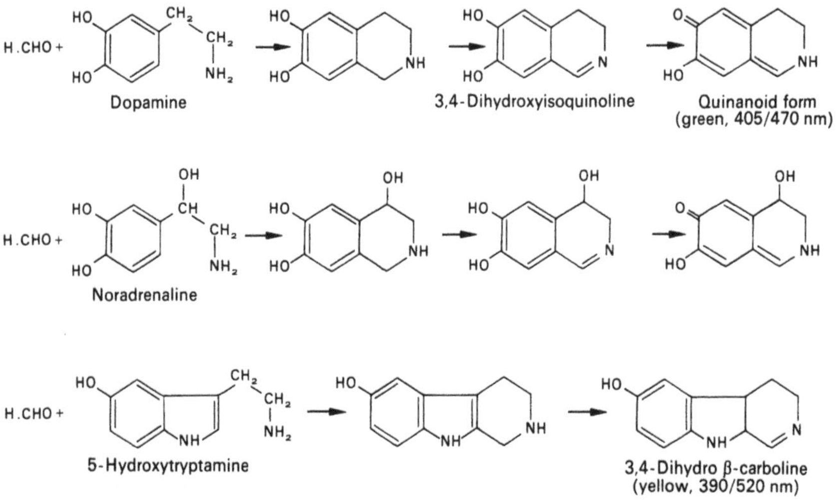

Dopamine

3,4-Dihydroxyisoquinoline

Quinanoid form (green, 405/470 nm)

Noradrenaline

5-Hydroxytryptamine

3,4-Dihydro β-carboline (yellow, 390/520 nm)

Fig. 3.5 Conversion of mono-amines to fluorescent derivatives.

tent of isolated vesicles indicate that each can contain only about 1600 molecules, at a concentration of about 0.25M, with a concentration in the whole nerve ending of some 0.03M. For 50 000 molecules to be there, a concentration of some 10M would be required. [If one takes the internal diameter of a cholinergic vesicle to be 500 Å, and assuming it is spherical, the internal volume is 6.6×10^{-17} ml. Since 1600 molecules are equivalent to 1.6×10^{-20} moles, the concentration is 0.25M. If, instead of 500 Å, one used a diameter of 400 Å, the concentration becomes 0.5M.] The number of 1600 molecules stored is similar to the number (1000 to 2000) calculated to react with the receptors to produce one miniature end plate potential [17]. We therefore seem to be faced with the apparent anomaly that 1000–2000 molecules are stored in each vesicle, that a similar number of molecules is required to produce a miniature end-plate potential, but that 50 000 molecules are released per quantum. Obviously the contents of more than one vesicle are released per quantum, but there is a further complication: there is reason to doubt that all of the acetylcholine contained within a vesicle is released at the same time. The results from studies on turnover rates of labelled acetylcholine indicate that it is only the 'hot', or recently-formed, acetylcholine that is released on stimulation and that the newly synthesised molecules represent only a small proportion of the total associated with the vesicles [18]. Much is still lacking therefore in our understanding of the quantitative relationships. However, if one reflects on the size of the vesicle (0.05μ in diameter) and the minute number of molecules involved, the progress to date seems truly remarkable!

3.4.3 Metabolism of acetylcholine

Acetylcholine is synthesised from choline and acetylcoenzyme A by the reaction catalysed by choline acetylase (Fig. 3.6) in the nerve endings

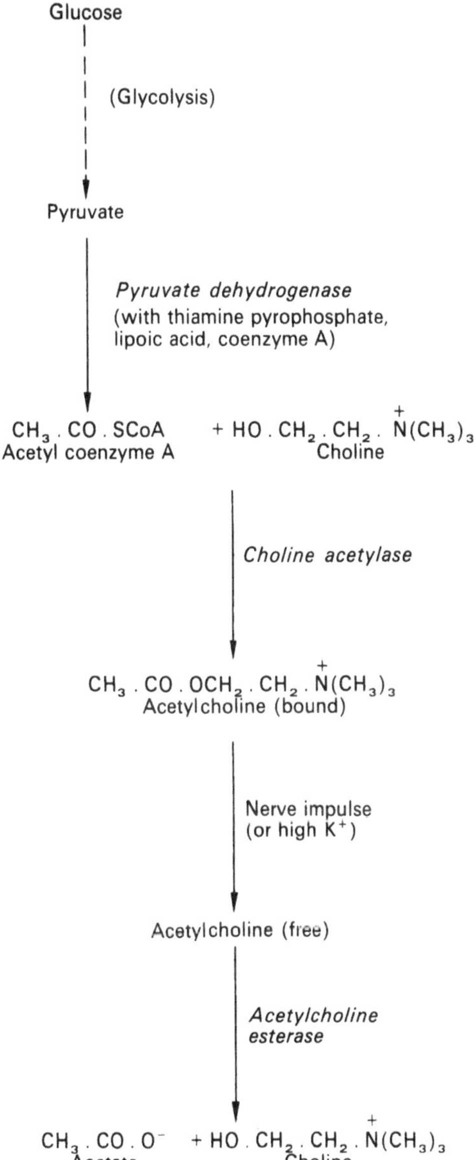

Fig. 3.6 Metabolism of acetylcholine.

of cholinergic systems. Since one of the precursors, acetylcoenzyme A, requires energy for its formation, the synthesis of the transmitter is an energy-consuming process. This reaction was studied by using eserine, an inhibitor of acetylcholinesterase, to prevent breakdown of the product of the reaction. This is essential because the esterase is widely occurring and extremely active (below).

The caudate nucleus has the most active choline acetylase in the central nervous system and produces acetylcholine at rates of some 10

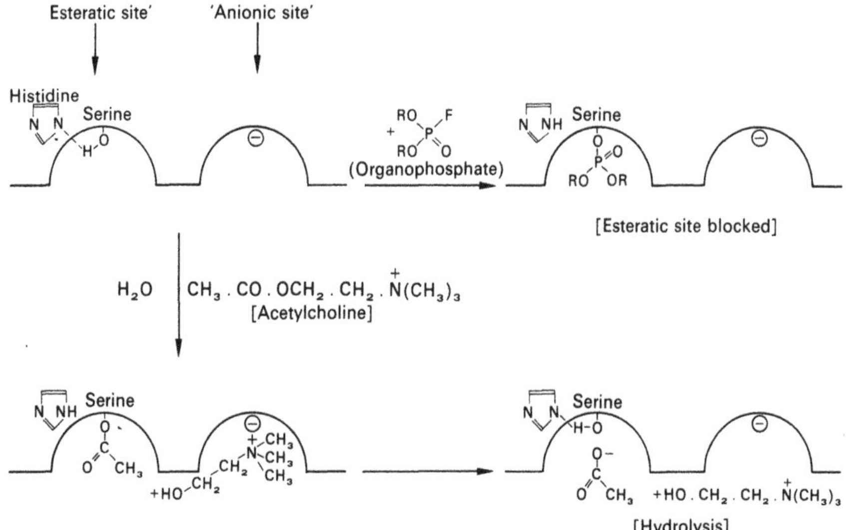

Fig. 3.7 Active site of acetylcholinesterase. The hydrolysis of the ester is inhibited by blockage of the esteratic site by organophosphate poisons.

to 15 μmoles per gram of fresh tissue per hour. This is much slower than the rates at which the transmitter can be destroyed (below). However such rates of synthesis, or the 2 to 3 μmoles/g/hr. produced in the cerebral cortex, are sufficient to maintain acetylcholine: as noted above, only a minute proportion of the stored transmitter is used normally *in vivo*, so the rate of required renewal will be relatively small. Synthesis requires a supply of choline which is transported into nerve endings more readily than is acetylcholine; Na^+ is also essential, probably due to its requirement for choline uptake and for acetylcholine storage. Hemicholinium blocks acetylcholine formation essentially by preventing choline uptake. The acetylcoenzyme A required to maintain these rates of acetylcholine synthesis was earlier thought to depend essentially on supplies of pyruvate from glycolysis, but realization of the reversibility of the choline acetyltransferase has thrown doubt on this. The results of studies on the turnover rates and kinetic properties of the enzyme from the electric organ of the Torpedo suggest that such formation of acetylcoenzyme A from acetylcholine (the enzyme requires only acetylcholine and coenzyme A for this) may be some 30 times more efficient than production of acetylcoenzyme A by acetylcoenzyme A synthase [19]. Such recycling of acetylcoenzyme A within the nerve ending cytoplasm is likely to occur when the concentration of acetylcholine exceeds its normal level of some 30m*M* there, in view of the K_m of the transferase for acetylcholine of 50m*M*.

Breakdown of this transmitter is considered to be the most efficient of all catabolic processes in the body: it is catalysed by perhaps the most active enzyme known to occur in nature. Hydrolysis of each molecule can occur in microseconds. The enzyme responsible, *acetylcholine esterase*, occurs very widely not only within but also without

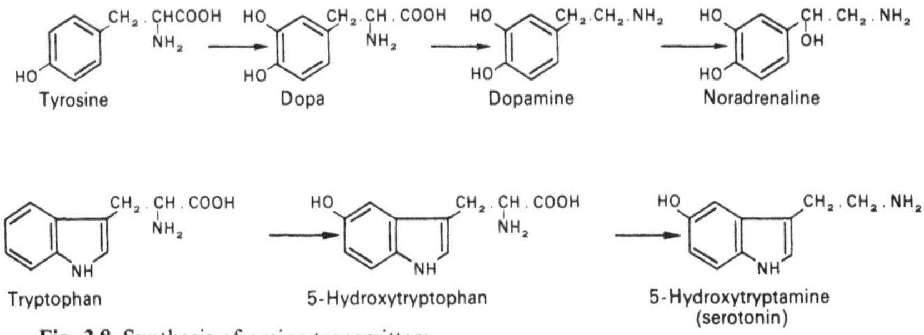

Fig. 3.8 Synthesis of amine transmitters.

the nervous system, presumably to ensure that 'leaked' transmitter is promptly destroyed. In nervous tissues it is most active in membrane fractions derived from the plasma membrane and the endoplasmic reticulum (the 'microsomes') and particularly in the membranes of the nerve endings. The reaction is hydrolysis of the ester to form choline and acetate (Fig. 3.6). The active centre of the esterase has been the subject of detailed studies and two parts have been recognised: an 'anionic' site which binds the quarternary nitrogen and an 'esteratic' site which contains a serine OH group. It is this group which is blocked by 'nerve gases', the organophosphate poisons, and is the reason for the toxicity of these agents (Fig. 3.7). Organophosphates react with many enzymes but it is cholinergic transmission which is particularly vulnerable due to the necessity of removing acetylcholine post-synaptically. Failure of respiration is the usual consequence arising from neuromuscular blockage in the diaphragm.

3.4.4 Catecholamines: noradrenaline and dopamine

These amines are synthesised from the amino acid, tyrosine, by the decarboxylation and hydroxylation reactions of Fig. 3.8; dopamine is an intermediate in the formation of noradrenaline. The first step, the hydroxylation of tyrosine, is catalysed by *tyrosine hydroxylase* and is the rate-limiting regulatory stage, being inhibited by the end-products. The next stage is removal of CO_2 by *Dopa decarboxylase* and like all amino acid decarboxylating enzymes, it requires pyridoxal phosphate as co-enzyme. Dopamine is converted enzymically to noradrenaline by oxidation of the side chain. The enzyme is *dopamine β-hydroxylase.*

These two 'hydroxylating' enzymes are of great interest: both require molecular O_2, so 'hydroxylase' seems a misnomer. They are really oxidases. Dopamine β-hydroxylase, a Cu^{2+}-protein, requires ascorbic acid as well as O_2 and is one of the few cerebral enzymes known to show a requirement for this chemical which is a vitamin in man and the guinea pig. Neither of these species can synthesise its own ascorbic acid, which in the guinea pig brain occurs in concentrations of 1–2 μmoles/g. Tyrosine hydroxylase, which requires O_2, Fe^{2+} and a pteridine cofactor, is found in the brain in the nerve endings, and as noted above, is the regulatory enzyme for the pathway. Dopamine β-hydroxylase is the only one of these hydroxylases so far detected

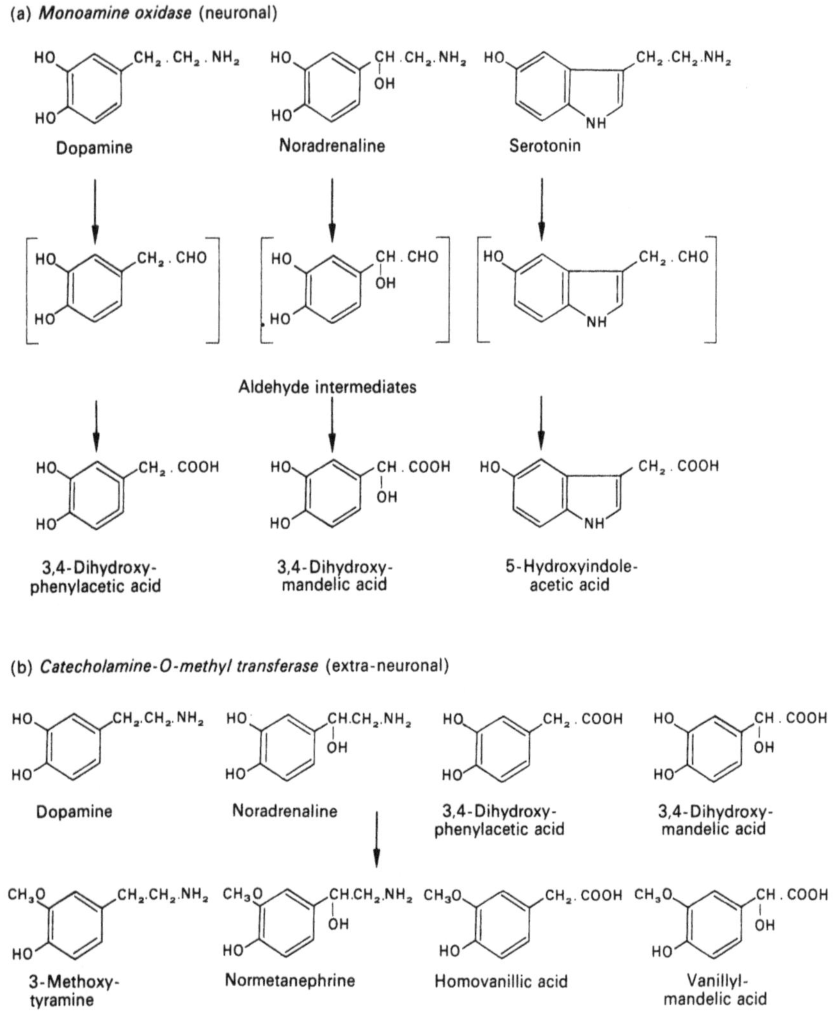

(a) *Monoamine oxidase* (neuronal)

Dopamine — Noradrenaline — Serotonin

↓

Aldehyde intermediates

↓

3,4-Dihydroxyphenylacetic acid — 3,4-Dihydroxymandelic acid — 5-Hydroxyindoleacetic acid

(b) *Catecholamine-O-methyl transferase* (extra-neuronal)

Dopamine — Noradrenaline — 3,4-Dihydroxyphenylacetic acid — 3,4-Dihydroxymandelic acid

↓

3-Methoxytyramine — Normetanephrine — Homovanillic acid — Vanillylmandelic acid

Fig. 3.9 Catabolism of amines.

within the synaptic vesicles; the others occur in the nerve endings, apparently in the extravesicular cytoplasm rather than inside the vesicles.

3.4.5 5-Hydroxytryptamine

This amine, commonly known as 'serotonin', is synthesised from tryptophan by similar series of reactions, hydroxylation and decarboxylation, to those producing the catecholamines. Tryptophan is oxidised by *tryptophan hydroxylase* to form 5-hydroxytryptophan (Fig. 3.8) which is then decarboxylated by a pyridoxal-phosphate requiring enzyme: *5-hydroxytryptophan decarboxylase*, which seems to be identical to Dopa decarboxylase.

Tryptophan hydroxylase, like the enzyme which oxidises tyrosine, also requires O_2, Fe^{2+} and a pteridine cofactor. Its activity in the brain is very low indeed and so it has proved difficult to detect and to characterise. It is distinct from tyrosine hydroxylase, and also from phenylalanine hydroxylase (which produces tyrosine), and is the rate-limiting enzyme for serotonin synthesis. Other amines (histamine and octopamine) are also now thought to possess centrally-active properties.

3.4.6. Breakdown of the biogenic amines

Fig. 3.9 gives the pathways for the catabolism of the amines: these are partly neuronal, catalysed by *mono-amine oxidases*, and partly extra-neuronal, catalysed by *catechol-O-methyl transferase*. The major excretion products are shown in Fig. 3.9. The mono-amine oxidases are of particular interest; they are mitochondrial in origin and can only be efficiently extracted from the mitochondria by treatment with detergents. Electrophoresis of partially purified extracts seemed originally to separate up to four forms ('isoenzymes'), but this is now believed to be due to varying proportions of phospholipid bound to the enzyme protein. It now appears to be agreed that two forms occur. Designated MAO_A and MAO_B, they differ in sensitivity to inhibitors and in substrate specificity: MAO_A acts preferentially on serotonin and noradrenaline, and is sensitive to inhibition by Clorgyline, whereas MAO_B prefers amines such as tryptamine, benzylamine and β-phenylethylamine. It is sensitive to inhibition by Pargyline and the tricyclic antidepressant drugs, Imipramine and Chlorpromazine. Both forms of the activity oxidise dopamine, although some species differences have emerged in that dopamine may be oxidised mainly by MAO_B in human brain and by MAO_A in rat brain [20, 21].

Much attention has been given to the amine transmitters due to the suspicion that they may be involved in psychiatric disorders. While there is little solid evidence to support the hypothesis that defects in amine metabolism may be involved in schizophrenia (the hypothesis is based largely on the hallucinogenic effects of drugs such as L.S.D. and mescaline, which bear some chemical resemblance to the biogenic amines), there is some circumstantial evidence that they may be of importance in depression. The basis for this is quite empirical – a group of antidepressive drugs was subsequently found to have striking inhibitory action on monoamine oxidases, the enzymes which destroy the amines (Fig. 3.9). Since that realisation, many drugs have been developed which act to modify endogenous amine concentrations, either by increasing or decreasing them. One difficulty is that most of these drugs are relatively unspecific in that, if they alter the level of one amine, they tend also to have similar effects on other amines. However, the results from studies on the effects of antidepressive drugs on individual forms of the oxidase give some hope that in the future specific drugs can be tailored to be effective against a potential enzymic abnormality in each individual case of depression. The amines are certainly involved in many aspects of mood and behaviour, as modifications in hypothalamic amine concentrations show.

3.4.7 Metabolism of the neuroactive amino acids

The two amino acids of Table 3.1 (γ-aminobutyrate and glycine) have been shown to exert inhibitory action on cerebral preparations; however they may also participate in general reactions of intermediary metabolism in the brain. γ-Aminobutyrate (Gaba) is formed by decarboxylation of glutamate, a dicarboxylic amino acid. The enzyme which is specifically responsible, *glutamate decarboxylase*, occurs enriched in nerve ending preparations (Fig. 3.4) as does Gaba itself. Gaba is removed by a mitochondrial transmination reaction; the succinic semialdehyde which results feeds back into the tricarboxylic acid cycle. Some 10% of the flux of carbon through this cycle normally passes through the 'Gaba shunt'. A further product of Gaba metabolism is γ-hydroxybutyrate, thought to be important in sleep.

Glycine, like glutamate and Gaba, is rapidly produced from glucose in the brain; its metabolism is closely linked to that of serine by enzymic reactions (such as the serine hydroxymethyl transferase) found throughout the body. Other amino acids which also have inhibitory action when applied to cerebral preparations include β-alanine and taurine; some are excitatory and include glutamate, aspartate, cysteine sulphinate and cysteine sulphonate (cysteate).

All of these amino acids can be regarded as putative neurotransmitters, but their widespread occurrence in relatively high concentration, together with their involvement often in more general aspects of cerebral metabolism, render difficult the assignment of a truly specific neurotransmitter function; some may be important as modulators of excitability [22].

3.4.8 The neuroactive peptides

'Substance P' (Fig. 3.10) was originally discovered in 1931 [23] in spinal cord and subsequently demonstrated in the brain (especially in the substantia nigra, hippocampus, globus pallidus and hypothalamus); though it is suspected of neurotransmitter function, it may act to modulate the neurotransmission of other agents [24, 25]. Interest in centrally-active peptides became acute on the discovery of the enkephalins and endorphins [26, 27] which showed morphine-like opioid properties. Morphine is the most effective analgesic known, but the dangers of addiction with the accompanying social consequences have stimulated the search for a non-addictive alternative analgesic. The discovery of endogenous compounds, which react with morphine receptor sites in the brain and which produce analgesia, fired hopes that the alternative had been found. Fig. 3.10 shows the amino acid sequence (91 residues) of β-lipotropin which contains one of the two enkephalins, methionine-enkephalin (sequence 61–65). The other is leucine-enkephalin, where leucine replaces the C-terminal

Substance P
Arg–Pro–Lys–Pro–Gln–Gln–Phe–Phe–Gly–Leu–Met

Neurotensin
Glu–Leu–Tyr–Glu–Asn–Lys–Pro–Arg–Arg–Pro–Tyr–Ile–Leu

β-Lipotropin (containing enkephalin and endorphin)
^{1}Glu–Leu–Thr–Gly–Glu–Arg–Leu–Glu–Gln–Ala–Arg–Gly–Pro–Glu–Ala^{15}
^{16}Gln–Ala–Glu–Ser–Ala–Ala–Ala–Arg–Ala–Glu–Leu–Glu–Tyr–Gly–Leu^{30}
^{31}Val–Ala–Glu–Ala–Glu–Ala–Ala–Glu–Lys–Lys–Asp–Ser–Gly–Pro–Tyr^{45}
^{46}Lys–Met–Glu–His–Phe–Arg–Trp–Gly–Ser–Pro–Pro–Lys–Asp–Lys–Arg^{60}
^{61}Tyr–Gly–Gly–Phe–Met–Thr–Ser–Glu–Lys–Ser–Gln–Thr–Pro–Leu–Val^{75}
^{76}Thr–Leu–Phe–Lys–Asn–Ala–Ile–Ile–Lys–Asn–Ala–His–Lys–Lys–Gly–Gln^{91}

Fig. 3.10 Some centrally-active peptides.

methionine. Another major opioid peptide is *β*-endorphin (sequence 61–69). The metabolic and functional relationships of the endorphins to the enkephalins are unclear: though studies on the regional distribution of their occurrence and of their receptor-binding affinities indicate that they occur and act independently, the topic remains controversial. Enkephalins are very labile in the brain *in vivo* due to peptidase activity; a stable analogue where the glycine of position 62 is replaced by D-alanine showed longer-lasting analgesic activity [28].

The list of centrally-active peptides grows almost daily, with N-terminal acetylaspartyl-, glutamyl-, *γ*-aminobutyryl-, *β*-alanyl- and histidyl-groups. One of these, *β*-alanyl-L-histidine (carnosine) is of particular interest in the olfactory system [29, 30]. The detection of some gastric hormones in the brain stimulated much immunochemical screening of cerebral preparations for related hormones; to date angiotensin II (8 amino acid residues), bombesin, bradykinin (9 residues), cholecystokinin (8 residues), gastrin (17 residues), somatostatin and VIP (28 residues) have all been detected using immunochemical techniques, though not all have been isolated and chemically identified. This is important because the use of radio-immunochemical methods, with the attendant dangers of cross-reactivity, may not provide sufficient proof of occurrence. Neurotensin, with 13 amino acid residues, occurs widely in the brain as well as in the ileum and jejunum [31, 32]. As noted above with respect to the amino acids, it is not clear whether many of these peptides act as modulators or transmitters, but they represent a fascinating group of centrally-active agents with many more almost certainly awaiting discovery.

3.5 Origin of synaptic vesicles

This is still an open question but two alternatives seem most likely: either they are formed in the perikaryon and migrate through the axon to their functional sites in the nerve ending (Section 3.7), or they are formed locally within the nerve ending. Although much of the earlier work was carried out on cholinergic vesicles, use of the fluorescent histochemical technique (Fig. 3.5) for catecholamines has given us a clearer picture of the occurrence of adrenergic vesicles. Some of the evidence is as follows:

(1) ligation of adrenergic nerves is followed by accumulation of noradrenaline above (proximal to) rather than below (distal to) the ligature;

(2) depletion of noradrenaline by reserpine is followed by reappearance of fluorescent material initially in the cell body;

(3) isotopically-labelled noradrenaline is transported through the axon in a proximodistal direction together with protein. The fastest rates of axoplasmic transport (Section 3.7) have been recorded for noradrenaline granules. However, whether the vesicles are produced in the cell body or in the nerve ending, the evidence currently available points clearly to the nerve endings as the sites of transmitter synthesis, though not necessarily within the synaptic vesicles. Most workers agree that acetylcholine is synthesised within the nerve endings but not in the synaptic vesicles, judging from the distribution of choline acetylase activity.

The same seems to be the case for Gaba: glutamate decarboxylase activity is thought to be present in nerve ending cytoplasm rather than in the vesicles. Like acetylcholine, it is the newly-synthesized Gaba which seems to be released on stimulation. In contrast, noradrenaline is thought to be synthesized within the synaptic vesicles, from the presence of dopamine β-hydroxylase there.

3.6 Post-synaptic events

The post-synaptic structure may be a motor end-plate, the membrane of a nerve cell body or a dendritic spine (Fig. 2.7). The dendrite is similar to the axon in one aspect: it has a resting potential, but it is thought to differ in that it may not be electrically excitable. It is, however, *chemically* excitable in that reaction of the transmitter with the receptor will produce a change in polarisation of the dendritic membrane. This is then conducted to the neurone which in turn will change in polarisation. The dendritic post-synaptic response (Fig. 3.11) has a lower amplitude and is of longer duration than the action potential. There seems to be no threshold, unlike the action potential, and it is not an 'all-or-none' process since it is clearly additive, i.e., repeated rapid release of transmitter may have an additive effect on increasing the amplitude of the post-synaptic potential (see the discussion of miniature end-plate potentials in Section 3.4.2).

The post-synaptic dendritic response may be one of depolarisation or of hyperpolarisation (Fig. 3.11): depolarisation, as described above, is when the potential changes in a positive direction from about -60mV towards zero or a positive value; hyperpolarisation is when the potential becomes more negative. In that case, since the potential of the cell is more negative, it is more resistant to depolarisation so that there is a tendency to inhibit excitation and the formation of an action potential. Depolarisation of the post-synaptic system is therefore excitatory and the resultant potential is known as the EPSP

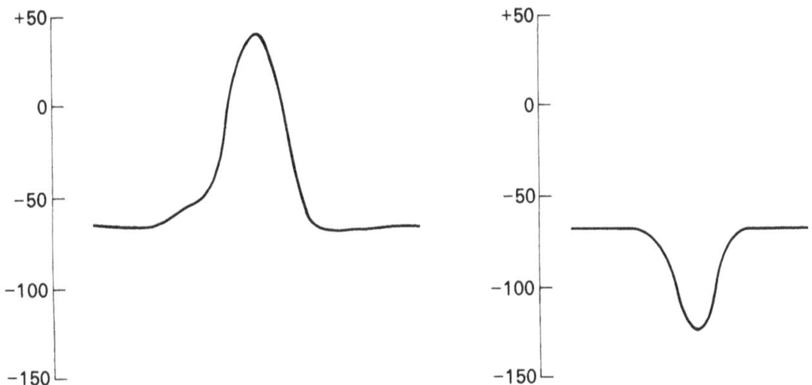

Fig. 3.11 Post-synaptic potentials. **(a)** An excitatory post-synaptic potential (EPSP) becomes from + 20 to + 40 mV, from the resting potential of about − 70 mV, a change of some + 100mV in depolarisation. **(b)** An inhibitory post-synaptic potential (IPSP) becomes about − 120 to − 140 mV, i.e. a change to a more negative potential in hyperpolarisation.

(excitatory post-synaptic potential). On the other hand, hyperpolarisation of the post-synaptic system is inhibitory and the potential is known as the IPSP (inhibitory post-synaptic potential). In excitatory synapses the transmitter acts to increase the permeability of the post-synaptic membrane to Na^+, causing depolarisation. In inhibitory synapses the transmitter acts to increase the permeability of the post-synaptic membrane to chloride and perhaps also to K^+, giving hyperpolarisation.

Whereas it is believed, on good evidence, that only one type of transmitter is stored and released from each individual nerve ending, some neurones, e.g. the Renshaw cells of the spinal cord, may have a multiplicity of types of synapse – in this case there seem to be two excitatory and one inhibitory. There is evidence that the input (A) to neurones of the sympathetic ganglia (Fig. 3.12) involves both nicotinic and muscarinic cholinergic receptors, with intermediation by interneuronal aminergic receptors. Here the responses to the inputs would be: fast excitatory at the nicotinic cholinergic receptor (i), slow excitatory at the muscarinic cholinergic receptor (ii), and slow inhibitory at the dopaminergic receptor (iii). The signal (B) from the ganglionic neuron will thus be modulated according to the balance of excitatory and inhibitory inputs [33]. This generalised, over-simplified picture shows only cholinergic and catecholaminergic events; all of the neurotransmitters discussed in this chapter are capable of such interactions. Further, there is recent evidence for pre-synaptic receptors, which may regulate more directly the release of neurotransmitters from the nerve ending. Some transmitters (e.g. acetylcholine in the cerebral cortex) may be excitatory or inhibitory, depending on the type of receptor. Others, such as the amino acids, glycine and γ-aminobutyrate, seem to be essentially inhibitory at the synapses that have been studied so far. For obvious reasons of accessibility

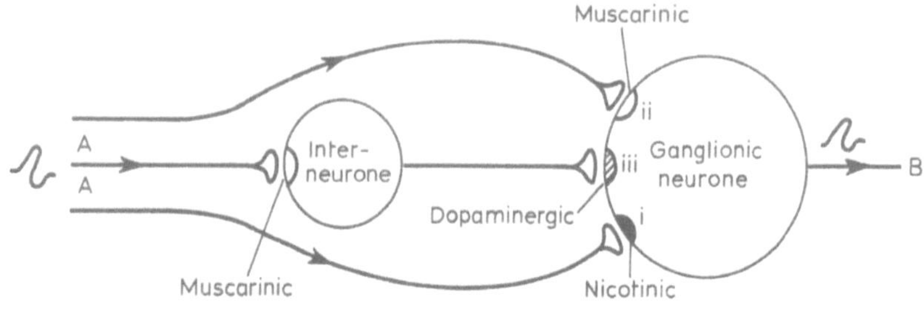

Fig. 3.12 Modulation of excitability in sympathetic neurones.

more work has been done on the spinal cord, the neuromuscular junction, and on peripheral nerves than on the central nervous system so far and it is still far from clear which pathways in the brain are mediated by which of the transmitters yet identified. It seems almost certain that there are transmitters still awaiting detection and identification. A diagrammatic representation of the major events occurring at cholinergic and adrenergic synapses is given in Fig. 3.13.

3.6.1 Involvement of cyclic nucleotides

Recent studies on various aspects of adrenergic transmission have raised the possibility of involvement of the cyclic adenine nucleotide, 3′, 5′-cyclic AMP, in subsequent metabolic events. Noradrenaline has been shown to activate *adenyl cyclase* (the enzyme which produces cyclic AMP from ATP) in inhibitory pathways in Purkinje cells of the cerebellum. Mediation by cyclic AMP of neurotransmission in sympathetic ganglia has also been suggested: in these ganglia, with preganglionic cholinergic nerve endings and adrenergic post-synaptic neurones, dopaminergic interneurones have also been suggested to be present (Fig. 3.12). According to the hypothesis, these dopaminergic cells receive signals from the cholinergic nerve endings and their nerve endings form synapses with the post-ganglionic neurones. The release of dopamine from the interneurones is suggested to activate adenyl cyclase in the post-ganglionic cell membrane, thus causing a local accumulation of cyclic AMP. This is then suggested to change the permeability of the post-synaptic cell membrane by causing phosphorylation of membrane proteins, which results in hyperpolarisation to render the post-ganglionic neurone less responsive to excitation. This hypothesis therefore invokes modulation of sympathetic transmission by cyclic AMP [33] and evidence in support is being sought. A direct link between synthesis of cyclic AMP and post-synaptic events has yet to be demonstrated. Moreover questions on the time sequence remain to be answered: it seems unlikely that the sequence of events postulated in the hypothesis (enzyme activation, diffusion of product to another enzyme, which is then activated to catalyse phosphorylation of membrane-bound proteins) can occur sufficiently quickly in relation to the rapid post-synaptic responses known to take place [34]. Certainly cyclic AMP is

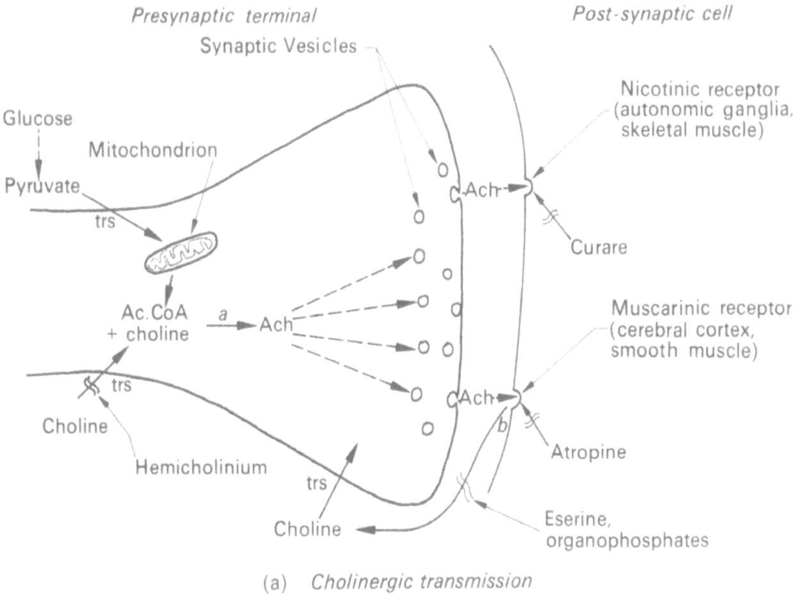

(a) *Cholinergic transmission*

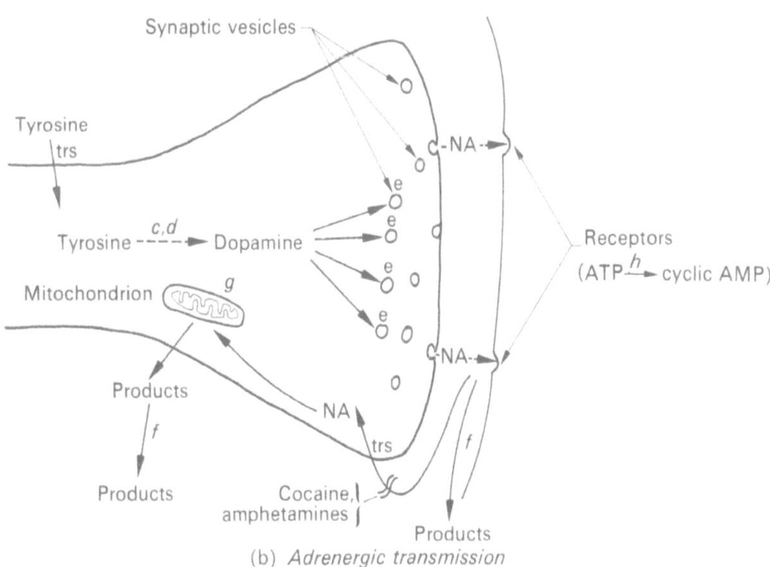

(b) *Adrenergic transmission*

Fig. 3.13 Schematic representation of neurotransmission. trs., transport, —//→ inhibition. **(a)** Cholinergic transmission. ACh, acetylcholine; a, choline acetylase; b, acetylcholine esterase (membrane bound). **(b)** Adrenergic transmission. NA, noradrenaline; c, tyrosine hydroxylase; d, Dopa decarboxylase; e, dopamine hydroxylase (vesicular); f, COMT; g, MAO; h, adenyl cyclase.

known to accumulate, the catecholamines are known to activate adenyl cyclase and the metabolic role most clearly eludicated for cyclic AMP in many systems, including micro-organisms, is activation of protein kinases. This was the basis of its discovery: its role in activating the protein kinase involved in glycogen phosphorylase activation [35], and relevant protein kinases have been described in neural tissues [36]. Cyclic AMP is also suspected of being involved with the functions of peptides, such as substance P, and with prostaglandins. Cyclic GMP seems intimately concerned with muscarinic cholinergic receptor function.

3.6.2 Receptors

While chemical analysis of neural membrane fractions has not shown any clear basis for the difference between excitable and non-excitable membranes, there is a wealth of pharmacological and physiological evidence for the presence of sites on the post-synaptic membrane which react specifically with the known chemical transmitters to cause rapid, transient and reversible changes in cation permeability. These receptor sites have mainly resisted attempts at isolation and purification so the evidence on the mechanisms of interaction with transmitters which cause conformational changes in the membrane macromolecules remains indirect. As noted above (Table 3.1), two types of cholinergic receptor have been detected pharmacologically: 'muscarinic', where transmission is mimicked by muscarine and blocked by atropine; 'nicotinic', mimicked by nicotine (Fig. 3.14) and blocked by curare. At present cholinergic receptors in the central nervous system seem to be essentially muscarinic. Both are distinct from the active site of cholinesterase. It was previously held by many workers that they were identical but recent work has eliminated this. The use of structural analogues of acetylcholine showed differences in specificity between enzyme and receptor; the two differed markedly in sensitivity to inhibitors (e.g. dithiothreitol). Also, denervated muscle showed increased sensitivity to acetylcholine but decreased enzymic activity. Some information on the nature of the receptor sites has also come from studies with local anaesthetics. Procaine (Fig. 3.14) interferes with interaction of acetylcholine at its receptor sites and so blocks transmission. If a butyl group is substituted on to the aromatic amino group, as in tetracaine, increased anaesthetic potency results, possibly due partly to increased penetration of the molecule to the receptor site.

Some of the approaches to identifying cholinergic receptors involve 'affinity-labelling' techniques whereby a stable, irreversible complex is formed between the receptor and an easily-identified chemical or biological label. A chemical approach was used by Karlin and his co-workers, making use of the disulphide groups known to occur near or at receptor sites. The free –SH groups, formed by reduction of the disulphides with dithiothreitol (Fig. 3.14) were complexed with substituted *N*-ethyl maleimides. The compound with the greatest

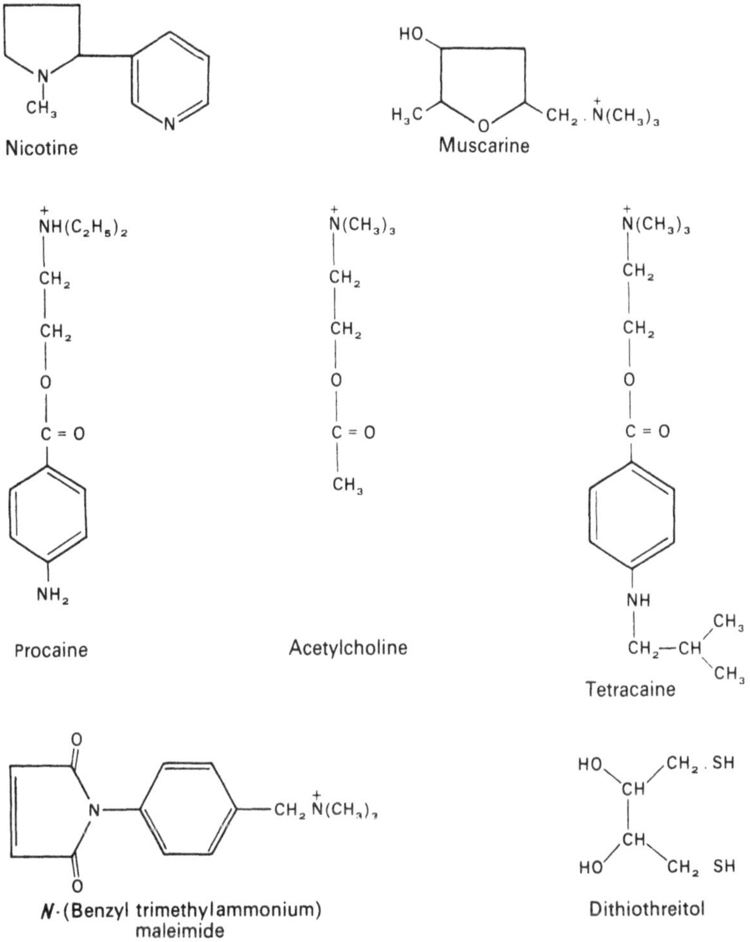

Nicotine

Muscarine

Procaine

Acetylcholine

Tetracaine

N-(Benzyl trimethylammonium) maleimide

Dithiothreitol

Fig. 3.14 Chemical structures.

affinity of those tested was *N*-benzyl trimethylammonium maleimide (Fig. 3.14). Using the tritiated compound, a protein complex of about 42 000 molecular weight was separated by gel electrophoresis from the electroplax organs of *Torpedo* fish, which contain cholinergic synapses with nicotinic receptors. The labelling of this protein was decreased in the presence of snake neuro-toxins [36]. The other main approach has been to label the membranes with radioactive neuro-toxins, such as α-bungarotoxin [37]. A protein of comparable molecular weight was then isolated. This nicotinic receptor protein has an apparent molecular weight of around 250 000 and in the electron microscope appears as a cluster of 6 units, each thought to be of about 40 000 molecular weight. Reaction of transmitter with this receptor somehow opens a sodium channel or 'ionophore'; the architectural relationship of an ionophore to the acetylcholine binding site is unknown [38]. It seems possible that the 'hole' which is seen in the centre of the receptor cluster (above) may represent the ionophore. It is at least feasible that a change in conformation of cluster units on reaction with acetylcholine could alter the size of the hole. Some hope of increasing our knowledge comes from indications that a toxin (histriotoxin) found in South American tree frogs interacts with the ionophore without detectable effects on acetylcholine binding. The treatments inherent in isolating and purifying protein membrane-components will almost certainly effect the native configuration of the protein so that it would be surprising indeed if its specialised biological function were retained. There is no doubt that drug-binding macro-molecules have been isolated and the lowering of the radioactivity of the isolated labelled protein in the presence of receptor-blocking agents argues for some degree of specificity. But the acid test is biological activity. This is a problem confronting many investigators working with nervous tissues. So many of the important functions reside in membrane-bound molecules which may only show their specific biological activity if the architectural integrity of their immediate environment is preserved.

The nicotinic cholinergic receptor is the best characterised to date, due essentially to its rich, homogeneous occurrence in electric organs; less progress has been made on isolation of cerebral receptors to other agents, although considerable pharmacological and immunochemical evidence for their occurrence is accumulating. Some caution should be exercised in assessing receptor identification from binding studies, either with pharmacological agents or with antisera. Uncertainties which may be present include pharmacological or immunological specificity of the agent employed, distinction between binding to receptor sites or to transport sites, the identity of the membrane fraction to which the ligand binds; the final emphasis must be on isolation and characterisation of the suspected receptor.

3.7 Neurone-axonal transport

The mechanism for replenishment of molecules, large and small, at nerve endings which may be many centimetres from their own cell bodies (Chapter 2) has been the subject of intense effort for a decade or more. Two possible processes may be involved: renewal by localised synthesis, or renewal by transport of pre-formed material synthesised in the nerve cell body. It has been clear for over 30 years that the material of the axoplasm moves in a proximodistal direction (i.e. in a direction away from the cell body towards the nerve ending). If a clamp or restriction is applied to a nerve so as to squash it, a build-up of material proximal to the 'bottleneck' can be seen under the light microscope (Fig. 3.15). Mitochondria accumulate here also and are less obvious on the distal side of the clamp. These facts, together with observations of higher enzyme concentrations (e.g. the mitochondrial enzyme, succinate dehydrogenase) on the proximal side, gave clear indications of a damming of the natural flow of materials down the axon [39].

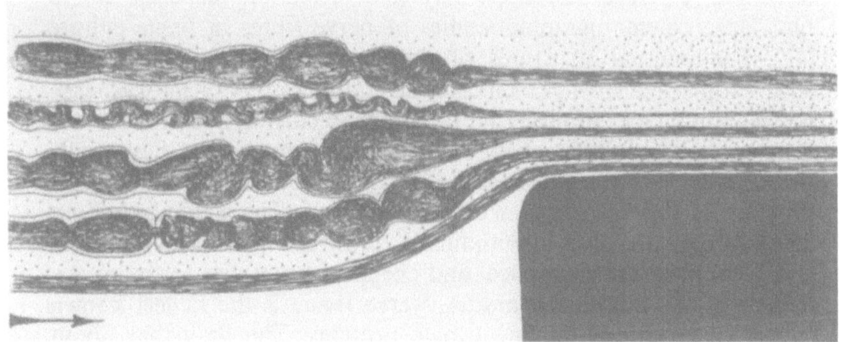

Fig. 3.15 Swelling of a nerve, proximal to a constriction. The black block represents the constriction to the nerve. To the left, on the proximal side (nearer to the cell body), the nerve becomes swollen within a few weeks and its contents present a beaded, convoluted appearance suggestive of damming of a stream, flowing from left to right [39].
Reproduced with the permission of Professor Paul Weiss, Rockefeller University, and J. Exptl. Zool.

Measurements of the rates of flow of a variety of molecules through the axon have been made using radio-isotopes. The general technique has been to inject the labelled compound into the cisterna magna or the spinal cord of laboratory animals, and subsequently to measure the rate of flow of the incorporated material through the appropriate nerve fibre. Estimation of radioactivity has been effected by direct counting in dissected segments of the nerve, or by autoradiography. The rate of flow has been found to vary to a certain extent according to the type of material being transported: two, and possibly three, types seem to occur. The slowest rates (1 to 2 mm/day) have been observed for proteins [40] labelled with tritiated leucine and for lipids labelled with ^{32}P-phosphate. Intermediate rates of 10 to 40 mm/day have also been noted using the same tracers. Really rapid rates of 200 to 500 mm/day [41] have been described for proteins and especially for catecholamine-containing granules [42]. The enzyme, choline acetylase, is apparently transported at the intermediate rate.

Many of the putative neurotransmitters, including peptides such as substance P, are now known to be transported in this way, as are many of the related synthetic enzymes. However it is not clear what proportion of the renewal of transmitters in nerve endings is due to flow and what to local synthesis. Although with rapid flow of enzymes it seems likely that most of the synthesis of neurotransmitters occurs at the nerve ending, there is some evidence for local synthesis of enzyme (below).

3.7.1 Mechanism of transport in axoplasmic flow

While little is known about this mechanism, flow due to constant pressure being exerted from the neuronal cell body seems unlikely and the kinetic properties of the process are incompatible with simple diffusion. A peristaltic process may be involved and has been seen in time-lapse cinematographic studies of nerve fibres in tissue culture [43]. The internal structures of the axon are certainly concerned with axoplasmic flow. Colchicine, an inhibitor of cell mitosis, suppresses some types of flow: the rapid type seems very sensitive to the drug and can be fully blocked, whereas the intermediate and especially the slow types are less affected. Colchicine binds specifically to the proteins of the *neurotubules* of the axoplasm. These tubules, about 250 Å in diameter, run longitudinally down the interior of the axon, and seem identical in physical and chemical properties to the *microtubules* of the mitotic apparatus. Nerve tissue is the richest known source of colchicine-binding tubule proteins. The brain has about 2 to 3 times the binding capacity per unit weight than the mitotic apparatus of various cells, and isolated axoplasm, about 10 times the capacity. Colchicine has been shown to inhibit tubule function by dissociating the tubular protein ('tubulin') into its inactive subunits. Dissociation is also caused by low temperature (3°C) which also inhibits axoplasmic flow. Tubulin is a protein of molecular weight about 120 000, consisting of two subunits. It is different from the globular protein (molecular weight 80 000) of another constituent of the axoplasm, the *neurofilaments*. These structures are smaller (80 to 100 Å) than the neurotubules, do not seem to be associated with axoplasmic flow, and do not bind colchicine. There is an apparent change during development: the immature brain contains a higher proportion of neurotubules than the mature brain, which contains relatively more neurofilaments. Relative activities of axoplasmic flow correlate well with this change. In the neurotubules, 1 molecule of tubulin dimer binds 1 molecule of colchicine and the binding requires Mg^{2+}. The nucleotide, GTP, is also required: 2 molecules are associated with each tubulin molecule. One of these is thought to stabilise the dimer and occupies a 'stable' site on the molecule. The second is more labile and its role is not clear. Interest in neurotubules and neurofilaments has been heightened by observations of 'neurofibrillary tangles' in the hippocampus and temporal lobe in ageing, and particularly in senile dementia. The precise relationship of the material of the tangles to these normally-occurring materials has not been resolved [44] but it seems possible that the tangles could contribute to cellular degeneration, perhaps by affecting axonal flow. While there is undoubted evidence for these very active and rapid transport mechanisms, not all material of the axons and nerve endings is renewed in this way.

3.7.2 Axonal protein synthesis

Independent protein synthesis has been shown to occur, not only in nerve endings, but also all along the axon. This has been most clearly demonstrated by studies on acetylcholinesterase activity, which can be irreversibly inhibited by organophosphate poisons. After such inactivation in peripheral nerves (mammalian sciatic and hypoglossal nerves, and in the axon of the Mauthner giant neurone of goldfish), the enzymic activity re-appears no more rapidly at sites near the neuronal cell body than it does at sites remote from the body. If the enzyme had been synthesised solely in the cell body, a gradation of renewed activity would have been expected, appearing first at proximal sites and subsequently in more distal segments of the nerve. In some experiments with paired mammalian nerves (e.g. the right and left hypoglossal nerves) both nerves were poisoned with organophosphate but only one was treated also with careful local applications of puromycin to inhibit any local protein synthesis [45]. Reappearance of enzymic activity was inhibited in the puromycin-treated nerve, but not in the untreated nerve. The results of the use of inhibitors are puzzling: the protein synthesis showed inhibition properties similar to those which occur in ribosomes rather than in the mitochondria also present within the axons. The RNA present in the axon, in preparations carefully stripped of surrounding myelin sheath and their glial cells, has been concluded (from its base-ratio analysis) to be of ribosomal type. Yet electron microscopists have failed to detect any ribosomes in the axons. There seems no doubt that a local axonal site for protein synthesis exists, but the mechanism of this synthesis is unclear at present.

Local sites in nerve endings for synthesis of enzymes have not been established but, as noted earlier in this Chapter, many of the enzymes required for transmitter synthesis have been detected there.

Protein therefore seems to be renewed at the nerve ending by a combination of local synthesis and transport of pre-formed protein; at present there is insufficient evidence to provide any basis for an assessment of the relative contribution or importance of each of the two processes. If significant amounts of protein do reach the nerve endings by transport it seems reasonable to expect active localised mechanisms for protein breakdown, but little knowledge on this is currently available.

In this Chapter we have been very briefly concerned with neuronal function and the interaction with endogenous chemicals intimately and directly involved; we can now turn to consider some aspects of the regulation of brain function by 'external' factors, either within the body external to the brain or belonging to the environment external to the body.

References

[1a] Hodgkin, A. L. and Katz, B. (1949), 'The effect of sodium ions on the electrical activity of the giant axon of the squid'. *J. Physiol.* **108**, 37–77.

[1b] Woodbury, J. W. (1965), 'Action potential: properties of excitable membranes'. *Neurophysiology* (Ruch, T. C., Patton, H. D., Woodbury, J. W. and Towe, A. L.), Chapter 2, London; Saunders.

[2] McIlwain, H. and Bachelard, H. S. (1971) *Biochemistry and the Central Nervous System* (4th. ed.), Churchill, London.

[3] Davies, M. (1973), *Functions of biological membranes.* Chapman and Hall, London.

[4] Glynn, I. M. (1968) 'Membrane adenosine triphosphatase and cation transport'. *Brit. Med. Bull.* **24**, 165–169.

[5] Skou, J. C. (1957), 'The influence of some cations on an adenosinetriphosphatase from peripheral nerves'. *Biochim. Biophys. Acta*, **23**, 394–401.

[6] Dunham, E. T. and Glynn, I. M. (1961) 'Adenosine triphosphatase activity and the active movement of alkali metal ions'. *J. Physiol.* **156**, 274–293. Whittam, R. H. (1962) 'The asymmetrical stimulation of a membrane adenosine triphosphatase in relation to active cation transport'. *Biochem. J.* **84**, 110–118.

[7] Kahlenberg, A., Galsworthy, P. R. and Hokin, L. E. (1968) 'Studies on the characterization of the sodium-potassium transport adenosinetriphosphatase'. *Arch. Biochem. Biophys.* **126**, 331–342.

[8] Hodgkin, A. L. (1958), 'Tonic movements and electrical activity in giant nerve fibres'. *Proc. Roy. Soc. B.* **148**, 1–37.

[9] Dale, H. H., Feldberg, W., and Vogt, M. (1936), 'Release of acetylcholine at voluntary motor nerve endings', *J. Physiol.* **86**, 353–380.

[10] Brown, G. L., Dale, H. H. and Feldberg, H. (1936), 'Reactions of the normal mammalian muscle to acetylcholine and eserine'. *J. Physiol.* **87**, 394–424.

[11a] Gray, E. G. and Whittaker, V. P. (1962), 'The isolation of nerve endings from brain'. *J. Anat. Lond.* **96**, 79–88.

[11b] DeRobertis, E., de Iraldi, A. P., de Lores Arnaiz, G. R. and Salganicof, L. (1962), 'Cholinergic and non-cholinergic nerve endings in rat brain'. *J. Neurochem.* **9**, 23–35.

[12] Marchbanks, R. M. (1967). 'The osmotically sensitive potassium and sodium compartments of synaptosomes'. *Biochem. J.* **104**, 148–157.

[13] Iversen, L. L. (1967), *The uptake and storage of noradrenaline in sympathetic nerves.* Cambridge University Press.

[14a] Falck, B. (1962), 'Observations on the possibility of the cellular localisation of monoamines by a fluorescent method'. *Acta Physiol. Scand.* **56**, suppl. 197.

[14b] Dahlström A. and Fuxe, K. (1964), 'Evidence for the existence of monoamine-containing neurones in the CNS'. *Acta Physiol. Scand.* **62**, suppl. 232.

[15] Katz. B. (1966), *Nerve, muscle and synapse.* New York, McGraw-Hill.

[16a] Whittaker, V. P. and Scheridan, M. N. (1965). 'The morphology and acetylcholine content of isolated cerebral cortical synaptic vesicles'. *J. Neurochem.* **12**, 363–372.

[16b] Wilson, W. S., Schultz, R. A. and Cooper, J. R. (1973), 'The isolation of cholinergic synaptic vesicles from bovine superior cervical ganglion and estimation of their acetylcholine content'. *J. Neurochem.* **20**, 659–667.

[17a] Katz, B. and Miledi, R. (1972), 'The statistical nature of the acetylcholine potential and its molecular components'. *J. Physiol.* **224**, 665–669.

[17b] Katz, B. and Miledi, R. (1973), 'The binding of acetylcholine to receptors and its removal from the synaptic cleft'. *J. Physiol.* **231**, 549–574.

[18] Marchbanks, R. M. and Israel, M. (1972), 'The heterogeneity of bound acetylcholine and synaptic vesicles'. *Biochem. J.* **129**, 1049–1061.

[19] Diebler, M.-F. and Morot-Gaudry, Y. (1977). 'Biosynthesis of acetylcoenzyme A in the electric organ of *Torpedo marmorata* in relation to acetylcholine metabolism'. *Biochem. J.*, **166**, 447–453.

[20] Achee, F. M., Gabay, S. and Tipton, K. F. (1977), 'Some aspects of monoamine oxidase activity in brain', *Progr. Neurobiol.*, **8**, 325–348.

[21] Johnston, J. P. (1968), 'Some observations upon a new inhibitor of monoamine oxidase in brain tissue', *Biochem. Pharmac.*, **17**, 1285–1297.

[22] Bachelard, H. S. (1981), 'Biochemistry of centrally active amino acids', In *Amino Acid Transmitters* (ed. Mandel, P. and DeFeudis, F. V.), Raven Press, Yew York.

[23] Von Euler, U. S. and Gaddum, J. H. (1931), 'An unidentified depressor substance in certain tissue extracts', *J. Physiol.*, **72**, 74–87.

[24] Malthe-Sørenssen, D., Cheney, D. L. and Costa, E. (1978), 'Modulation of acetylcholine metabolism in the hippocampal cholinergic pathway by intraseptally injected substance P', *J. Pharmac. Exp. Ther.*, **206**, 21–28.

[25] Henry, J. L., Krnjević, K. and Morris, M. E. (1975), 'Substance P and spinal neurons', *Can. J. Physiol. Pharmac.*, **53**, 423–432.

[26] Hughes, J., Smith, T. W., Kosterlitz, H. W., Fothergill, L. A., Morgan, B. A. and Harris, H. R. (1975), 'Identification of methionine-enkephalin structure', *Nature*, **258**, 577–579.

[27] Bradbury, A. F., Smyth, D. G. and Snell, C. R. (1976), 'Lipotropin: precursor of two biologically-active peptides', *Biochem. Biophys. Res. Commun.*, **69**, 950–956; Prohormones of beta-melanotropin and corticotropin: structure and activation, *CIBA Found. Sympos.*, **41**, 61–75.

[28] Pert, A. (1976), 'Behavioral pharmacology of D-alanine 2-methionine enkephalin amide and other long-acting opiate peptides', In: *Opiates and Endogenous Opioid Peptides* (ed. Kosterlitz, H. W.), p. 87–94, North-Holland, Amsterdam.

[29] Emson, P. C., Hunt, S. P., Gilbert, R. T. F., Wu, J. Y., Rehfeld, J. F. and Fahrenkrug, J. (1980), 'Current knowledge of the storage and release of transmitter candidate amino acids and neuropeptides', In: *Synaptic Constituents in Health and Disease* (eds. Brzin, M., Sket, D. and Bachelard, H.), pp. 57–80, Pergamon, Oxford.

[30] Harding, J. and Margolis, F. L. (1976), 'Denervation in the primary olfactory pathway of mice. III. Effects on enzymes of carnosine metabolism', *Brain Res.*, **110**, 351–360.

[31] Carraway, R. and Leeman, S. E. (1976), 'Characterization of radioimmunoassayable neurotensin in the rat', *J. Biol. Chem.*, **251**, 7045–7052.

[32] Polak, J. M., Sullivan, S. N., Bloom, S. R., Buchan, A. M. J., Facer, P., Brown, M. R. and Pearse, A. G. E. (1977), 'Specific localization of neurotensin to the N cell in human intestine by radioimmunoassay and immunocytochemistry', *Nature*, **270**, 183–184.

[33] Greengard, P. (1976), 'Possible role for cyclic nucleotidës and phosphorylated membrane proteins in postsynaptic actions of neurotransmitters', *Nature*, **260**, 101–108.

[34] Rodnight, R. (1979), 'Cyclic nucleotides as second messengers in synaptic transmission', In: *Physiological and Pharmacological Biochemistry* (ed. Tipton, K. F.), **26**, 1–80, Univ. Park Press, Baltimore.

[35] Sutherland, E. W., and Rall, R. W. (1960), 'The relation of adenosine-3′, 5′-phosphate and phosphorylase to the actions of catecholamines and other hormones'. *Pharmacol. Revs.* **12**, 265–299.

[36] Keitler, M. J., Cowburn, D. A., Prives, J. M. and Karlin, A. (1972), 'Affinity labelling of the acetylcholine receptor in the electroplax'. *Proc. Nat. Acad. Sci. U.S.* **69**, 1168–1172.

[37a] Miledi, R., Molinoff, P. and Potter, L. T. (1971), 'Isolation of the cholinergic receptor protein of Torpedo electric tissue'. *Nature*, **229**, 554–557.

[37b] Meunier, J-C, Olsen, R., Menez, A., Morgat, J-L., Fromageot, P., Ronse-

ray, A. M., Boquet, P. and Changeux, J-P. (1971), 'Quelques propriétés physiques de la protéine réceptrice de l'acétylcholine étudiées à l'acide neurotoxine radioactive'. *CR. Acad. Sci. Paris*, **273D**, 595–598.
[38] Heidmann, T. and Changeux, J.-P. (1978), 'Structural and functional properties of the acetylcholine receptor protein in its purified and membrane-bound states', *Ann. Rev. Biochem*, **47**, 317–357.
[39] Weiss, P. and Hiscoe, H. B. (1948), 'Experiments on the mechanism of nerve growth'. *J. Exptl. Zool.* **107**, 315–395.
[40] Taylor, A. C. and Weiss, P. (1965), 'Demonstration of axonal flow by the movement of tritium-labelled protein in mature optic nerve fibres'. *Proc. nat. Acad, Sci. U.S.* **54**, 1521–1527.
[41] Ochs, S., Sabri, M. I. and Johnson, J. (1969). 'Fast transport system of materials in mammalian nerve fibres'. *Science*, **163**, 686–687.
[42] Dahlström A. and Haggendal, J. (1966), 'Studies on the transport and life-span of amine storage granules in a peripheral adrenergic neuron system'. *Acta Physiol. Scand.* **67**, 278–288.
[43] Weiss, P. (1958), 'The concept of perpetual neuronal growth and proximo-distal substance convection', in *Regional Neurochemistry* (ed. Kety, S. S. & Elkes, J.), Pergamon, London, p. 220–242.
[44] Katzman, R., Terry, R. D. and Bick, K. L. (eds), (1978) *Alzheimer's Disease*, Aging, Vol. 7, Raven Press, New York.
[45] Koenig, E. (1969), 'Nucleic acid and protein metabolism of the axon'. *Handbook of Neurochemistry*, **2**, 423–434 (ed. Lajtha, A.), Plenum, N. Y.

4 Adaptive processes in the brain

4.1 Inducible enzymes

Although much has been known for a decade or more about enzyme induction in micro-organisms and in mammalian organs such as the liver, analogous processes have only recently been detected in neural tissues. In a review on neural plasticity some years ago, a statement [1] that nothing was known about enzyme induction in nerve cells, could not easily be challenged. However, in the interval many examples have been described, and these mainly concern enzymes involved in transmitter metabolism or function.

The preceding chapter of this book included a brief description of tyrosine hydroxylase as the rate limiting stage of catecholamine synthesis and there now is evidence that this enzyme is inducible. Release of noradrenaline on stimulation of adrenergic nerves is followed immediately by a rise in the rate of its synthesis from tyrosine. This was believed to be due essentially to the decrease in the end-product inhibition of tyrosine hydroxylase, normally by noradrenaline. However subsequent work indicated that the activity of tyrosine hydroxylase *in vitro* also increased. If the increase in rate of synthesis of the catecholamines *in vivo* had been due solely to de-inhibition of the enzyme, a prolonged increase in its activity *in vitro* would not be expected and the possibility therefore arises of increased enzyme protein formation in addition to activation of the pre-existing enzyme. Further evidence that the increase in tyrosine hydroxylase activity of adrenals and superior cervical ganglia could be prevented by inhibitors of protein synthesis (e.g. cycloheximide) tended to support the growing view that an induction of enzyme synthesis was involved [2]. Formation of the enzyme is thought to occur in the neuronal cell body from where it is transported to its functional site in the nerve ending, by the neurone-axonal flow mechanisms described in Chapter 3. Induction of tyrosine hydroxylase has been examined closely in adrenergic neurones of ganglia from the sympathetic nervous system. Treatment with reserpine, which releases amines from their storage sites (Chapter 3), is followed by increased tyrosine hydroxylase activity; lowering the temperature, which is associated with increased adrenergic activity, also results in increased enzymic activity [3]. Furthermore, the increase in the number of ganglionic synapses which occurs during development is paralleled by an increase in hydroxylase activity; if the pre-ganglionic nerve trunk is cut, the increased enzymic activity in post-synaptic neurones is prevented [4]. These observations are suggested to be consistent with *trans-synaptic* regulation: i.e. that events at presynaptic nerve terminals (release of neurotransmitters) regulate enzymic levels in the post-synaptic cells. Further evidence has come from the use of high K^+ concentrations to depolarise cultured sympathetic ganglia: again tyrosine hydroxylase activity increased [5]. Similar results were obtained with dibutyryl cyclic AMP (Chapter 3). In the experiments with high K^+, the effect was enhanced by the presence of theophylline (a caffeine-like drug which prevents break-down of 3′, 5′-cyclic AMP). There is some evidence, from the use of inhibitors, that induction of synthesis of new enzyme is involved. The results with cyclic AMP are also compatible with a post-synaptic process (Chapter 3). Since the ganglia used in such studies are innervated by cholinergic nerves, it seems likely that it is the pre-synaptic release of acetylcholine that causes the trans-synaptic increase of post-synaptic enzyme [2].

This therefore provides a good example of metabolic adaptation, i.e. of enzyme induction, neuronally mediated in response to excitation and neurotransmission. While the mechanism of this adaptation remains unclear, it seems unlikely to be due to induction by the substrate, tyrosine, in the absence of any evidence that the concentration of tyrosine increases. There is a strong possibility, therefore, that the enzyme is induced directly by neurohumoural agents, where presynaptic release of a neurotransmitter is followed by enzyme synthesis in postsynaptic cells. This seems clearly to be the case in the adrenal medulla and sympathetic ganglia, where immunoassay provided further evidence for the presence of increased amounts of enzyme protein. However in certain brain regions studied (hippocampus, hypothalamus and locus coeruleus) the evidence is less firm for synthesis of new protein: while tyrosine hydroxylase activity was increased under the conditions described above (including treatment with reserpine and cholinergic stimulation), immunoassay failed to show a significant increase in enzyme protein [6].

Examples are known of induction of cerebral enzymes by their specific substrates and there is growing knowledge on adaptive processes also involving hormones, drugs and environmental stimuli. Some of these are described in the following sections.

4.1.1 Adaptation to specific substrates

The first example of this is the response of β-hydroxybutyrate dehydrogenase to the presence of substrate in the blood stream. It was originally conceived as a direct induction of the enzyme by its substrate but the results of more recent work have thrown doubts on this: the presence of substrate is now believed to result in the maintenance of existing enzyme, rather than in induction of synthesis of new enzyme. The detection of this response of the cerebral dehydrogenase came almost simultaneously from independent studies on rat and human brain, based on the use of quite different techniques. Both were accidental findings. Working with rats, Sokoloff and his co-workers had become interested in the effects of thyroxin on the developing brain, in particular its effects on the *in vitro* incorporation of amino acids into mitochondrial protein. In such studies, oxidisable substrates (such as succinate) are presented to the incubated mitochondria to ensure adequate endogenous supplies of ATP. One of these, β-hydroxybutyrate, was found to support the amino acid incorporation in the mitochondria from immature rat brain far more effectively than in mature mitochondria, due to the efficiency of ATP formation (Fig. 4.1). Further investigation confirmed that the immature mitochondria were relatively rich in β-hydroxybutyrate dehydrogenase activity (Figs. 4.1, 4.2) whereas the cerebral mitochondria from mature animals contained very little [7]. Comparisons of the changes in mitochondrial enzyme activities during development showed that, whereas enzymes of the tricarboxylic acid cycle or of the respiratory chain (e.g. cytochrome oxidase, Fig. 4.1.) increased over the first few weeks of life to reach a stable level, retained in the adult, the β-hydroxybutyrate dehydrogenase activity rose in the first few weeks, but fell immediately after the young rats were weaned. Prolongation of weaning resulted in a delay in the fall of the dehydrogenase activity and analysis of the maternal rat milk revealed that it was rich in ketone bodies (β-hydroxybutyrate and acetoacetate). The conclusion was that the presence of the substrate had induced enzymic activity which disappeared when the substrate was no longer present. However, attempts to re-induce the activity in adult rats by starvation (when the β-hydroxybutyrate concentration in the blood

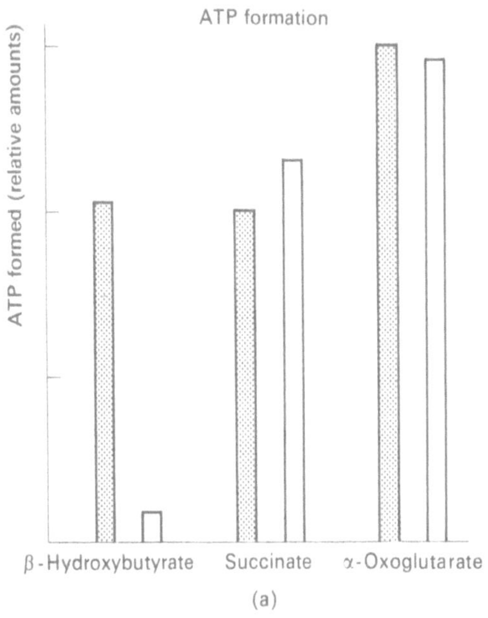

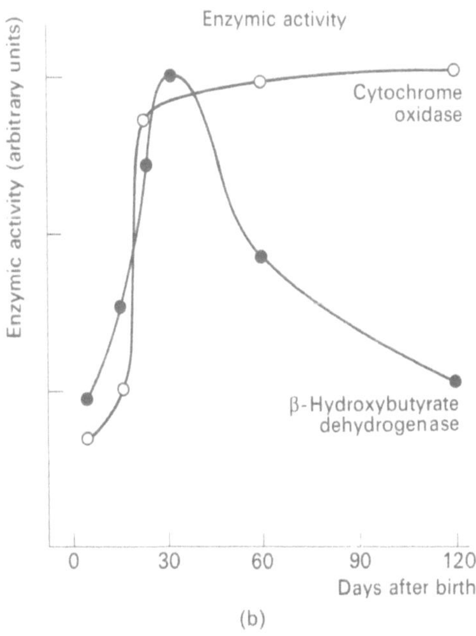

Fig. 4.1 Metabolism of β-hydroxybutyrate in young and adult rats. **(a)**: ATP formation in the mitochondria occurs at similar rates in the presence of succinate or α-oxoglutarate. When β-hydroxy-butyrate is the substrate, mitochondria from the adult (clear column) produce less ATP than from the young animals (dark column). **(b)**: β-Hydroxybutyrate dehydrogenase falls after weaning at about 4 weeks of age, whereas cytochrome oxidase activities of the same mitochondrial preparations remain constant [7].

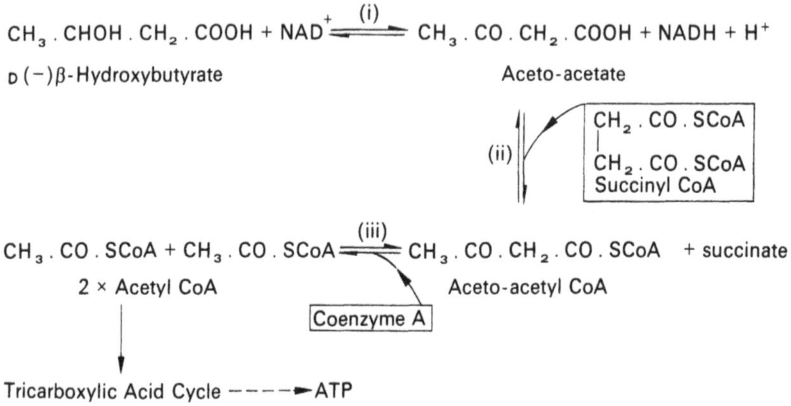

$$CH_3 . CHOH . CH_2 . COOH + NAD^+ \xrightleftharpoons{(i)} CH_3 . CO . CH_2 . COOH + NADH + H^+$$

D (−)β-Hydroxybutyrate — Aceto-acetate

(ii) ⇅ with Succinyl CoA ($CH_2 . CO . SCoA$ / $CH_2 . CO . SCoA$)

$$CH_3 . CO . SCoA + CH_3 . CO . SCoA \xrightleftharpoons{(iii)} CH_3 . CO . CH_2 . CO . SCoA + \text{succinate}$$

2 × Acetyl CoA — Coenzyme A — Aceto-acetyl CoA

2 × Acetyl CoA → Tricarboxylic Acid Cycle ----→ ATP

(i) β-Hydroxybutyrate dehydrogenase (EC 1.1.1.30)
(ii) 3-Oxo acid-coenzyme A transferase (EC 2.8.3.5)
(iii) Acetoacetyl-coenzyme A thiolase (EC 2.3.1.9)

Fig. 4.2 Oxidation of ketone bodies.

stream becomes elevated as a result of mobilisation of depot fats) proved unsuccessful. The net utilisation of the β-hydroxybutyrate by the brain increased, but the amounts of available dehydrogenase activity remained unchanged; the enzyme under fed conditions was presumably far from being saturated with substrate. So it was argued that in the adult, the enzyme normally present, though much less than in the young animal, remained constant and that the increased utilisation of ketone bodies on starvation was due to increased availability of substrate rather than to induction of enzymic activity [8]. Further support for this conclusion was given by observations of increased cerebral utilisation of acetoacetate in adult rats after acute infusion of the acetoacetate [9].

In the same year as Sokoloff's initial experiments on the rat, a similar type of adaptation was reported to occur in man. Prolonged fasting of obese patients had been carried out under clinical conditions which allowed sampling of arterial and venous blood. After 5 to 6 weeks' starvation on a diet which contained only water, salt, vitamins and flavouring, their mental ability was unimpaired, yet the arterio-venous difference was such that insufficient glucose to maintain consciousness was apparently being used by the brain. The respiratory quotient (from measurements of O_2 and CO_2) was only 0.63 compared with the normal value of close to one. Oxygen consumption was in the normal range, so some substrate other than glucose must have been consumed. Analysis of the arterial and venous blood showed that β-hydroxybutyrate was being consumed at a rate significantly greater than that of glucose (Fig. 4.3). It is well established that under normal conditions the mammalian brain cannot replace glucose as its major energy source. Few substrates are effective in reversing hypoglycaemic coma and those that are capable of this are thought to be converted to glucose elsewhere in the body before utilisation in the brain. Yet here

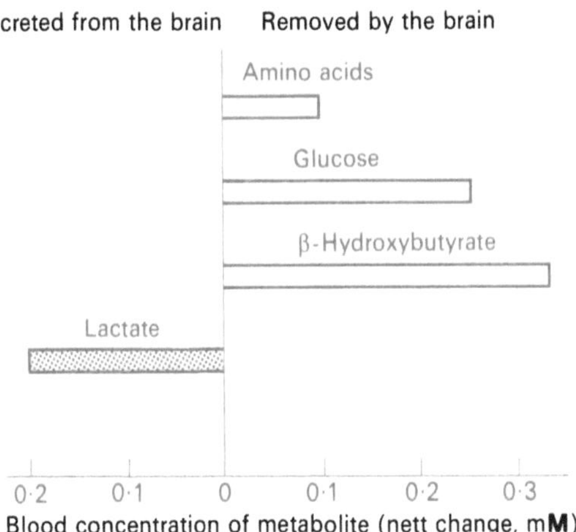

Fig. 4.3 Cerebral arterio-venous differences during starvation [10].

was a condition in which the human brain was clearly capable of consuming an alternative to glucose, β-hydroxybutyrate [10].

At present it is not clear if similar mechanisms operate in the adaptation to circulating ketone bodies in man and the rat because comparable experiments have not been performed on both species. We do not know the time course, after beginning the fasting of obese humans, of the rise in blood concentration of β-hydroxybutyrate and whether the cerebral dehydrogenase activity remains constant or rises during the period of starvation. For the moment it is assumed that the situation is similar; i.e., that in man, the enzymic activity remains constant and that the increased consumption of β-hydroxybutyrate reflects its increased concentration in the circulation. We therefore are forced to think in terms of two processes: the first, a form of adaptation, is the change in enzyme content of cerebral mitochondria during development in which the amount of enzyme present does appear to be conditioned by the substrate available, and which does not seem to occur in the adult brain. The second is the increased rate of utilisation of the substrate due to its increased availability, and is therefore not really an adaptive process since it might be expected to occur generally with any enzyme system whose substrate changes from sub-saturating to saturating concentrations. An alternative form of adaptation which might be present is in the transport of the ketone body from the blood to the brain. This is not possible to assess in the human starvation experiments without the appropriate time courses of change in blood levels and arterio-venous difference in the β-hydroxybutyrate; prolonged starvation of rats has not proved to result in increased enzymic activity but could result in increased rates of transport. In rat brain slices incubated *in vitro*, preparations from immature animals oxidised acetoacetate more rapidly than those from mature animals so the possibility of changes in the capacity for transport of ketone bodies can now be studied in immature and mature rat brain *in vitro* [11].

Subsequent studies *in vivo* have indeed indicated that the *capacity* of the transport system is increased on starvation; the investigators observed an increase in V with no change in K_m, and suggested an induced increase of carrier protein had occurred [12]. This remains to be confirmed, which might prove difficult. While the use of inhibitors of protein synthesis can provide circumstantial evidence for induction, firm evidence rests on proof of increased amounts of protein, usually by isolation or by immunochemical means which depend on isolation and purification. However while evidence for the presence of 'carrier' proteins in cerebral systems is sound, none has been isolated to date. The changes in β-hydroxybutyrate and acetoacetate concentrations, especially in the case of the immature rats, are essentially dietary changes; the example which follows involves regulation of the available amounts of a cerebral enzyme with substrate induction and product repression.

Glutamate decarboxylase catalyses the formation of γ-aminobutyrate (GABA) from glutamate (Chapter 3). Both GABA and the decarboxylase are mainly neural in occurrence, little of either being found elsewhere in the mammalian body. The functional importance of the decarboxylase in the brain is that it helps to regulate the relative proportions of glutamate (an excitant amino acid) and GABA (an inhibitory amino acid, almost certainly an inhibitory neurotransmitter; Chapter 3). The enzyme has been shown to occur in the nerve endings. The activity of mouse brain glutamate decarboxylase, measured *in vitro*, was almost doubled some 4hr. after an intraperitoneal injection of its substrate, L-glutamate. During this period the concentration of glutamate in the brain increased by less than 50% [13]. Other related amino acids (L-aspartate, DL-glutamine, GABA) or cortisol were ineffective. If the animals were pretreated with an inhibitor of protein synthesis, actinomycin D (2 mg/kg body weight), the effect of glutamate was diminished. The possibility that the increase in enzymic activity had been due to substrate-protection of a labile enzyme was considered unlikely since the slow decrease in enzymic activity of incubated brain slices was not affected by the presence of glutamate in the incubation medium [13]. It seems plausible, therefore, to consider that the increase in enzymic activity was due to substrate-induced synthesis of new enzyme. These results are slightly surprising in that many workers have found that glutamate cannot easily be caused to accumulate in the brain although it exchanges between blood and brain relatively readily (see Chapter 2).

Synthesis of this enzyme may also be subject to suppression by the product; *in vivo* accumulation of GABA is followed by diminished enzymic activity. GABA also is not readily taken up from the bloodstream to the brain so in these experiments an inhibitor of GABA catabolism was used to effect its accumulation. A subcutaneous injection of amino-oxyacetic acid into young mice caused a 5-fold increase in brain GABA content within 6hr. which was followed during the course of one day by a decrease in glutamate decarboxylase activity

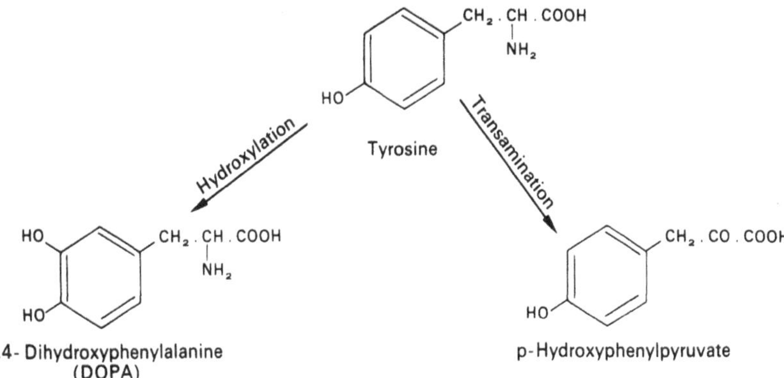

Fig. 4.4 Alternative pathways of tyrosine metabolism.

[14]. The use of an inhibitor to produce accumulation of an endogenous chemical presents problems not faced if the accumulation could be achieved more directly by elevating its concentration in the bloodstream, so various potential interpretations alternative to that of supression of enzyme synthesis required assessment. Glutamate decarboxylase is a pyridoxal phosphate requiring enzyme [15]; indeed, apart from pyridoxal kinase, it is the cerebral enzyme most sensitive to reagents which interfere with binding of pyridoxal phosphate. Moreover, like all transaminases, the enzyme which removes GABA and which in these experiments was inhibited to cause its accumulation, GABA-α-oxoglutarate transaminase, also requires pyridoxal phosphate. Pretreatment with pyridoxal phosphate was found to reverse the effect of amino-oxyacetic acid in that the decrease in decarboxylase activity was prevented. Other pyridoxine antagonists were tested—those which caused no change in GABA concentration (thiosemicarbazide and hydroxylamine) had no effect on the decarboxylase activity, whereas hydrazine, like amino-oxyacetic acid, caused both the increase in GABA and the decreased enzymic activity. The diminution of activity is not likely to be due to direct product inhibition of decarboxylase by GABA because it does not inhibit the enzyme *in vitro*; the results seem therefore to be consistent with suppression of enzyme synthesis by GABA. The adaptive processes which affect the enzyme which produces GABA may thus provide a potentially effective and sensitive means of regulating presynaptic GABA levels.

4.1.2 Adaptation to the product of an alternate pathway

A fascinating example of the product of one enzyme causing an adaptive response in an enzyme of an alternate pathway in the brain is given by the stimulation of tyrosine transaminase by 3, 4-dihydroxyphenylalanine (Dopa), which is produced by tyrosine hydroxylase (Fig. 4.4; see also Chapter 3). Brain biogenic amine concentrations can be decreased by administration of reserpine, which depletes catecholamines and serotonin (Chapter 3) by preventing their storage; treat-

Table 4.1 Effect of endogenous L-3, 4-dihydroxy-phenylalanine (Dopa) on cerebral tyrosine transaminase activity [16].

Treatment	*Enzymic activity (% of untreated)*
α-Methyltyrosine	60
α-Methyltyrosine + Dopa	110
Reserpine	65
Reserpine + Dopa	105

ment with α-methyltyrosine acts more specifically to decrease catecholamine concentrations by inhibiting tyrosine hydroxylase, but not tryptophan hydroxylase. Treatment of animals for some hours with either of these reagents caused a marked decrease in the cerebral tyrosine transaminase activity [16]. This was reversed by pretreatment with Dopa (Table 4.1). Presumably the transaminase is not directly affected normally (i.e. in the untreated animals) by Dopa since the enzymic activity measurements were performed on tissue extracts *in vitro*, so it seems possible that the presence of Dopa induces formation of the active transaminase enzyme. This is not certain, however, especially without directly testing the effects of Dopa on the isolated transaminase and without some assessment of the effects of inhibitors of protein synthesis. Normally cerebral transaminase capacity is some 100 times that of the hydroxylase, as measured *in vitro*, so that relatively little tyrosine would be available for Dopa formation. However, if the endogenous concentrations of Dopa should fall, the decreased transaminase activity should render more tyrosine available for Dopa formation. It seems possible, therefore, that Dopa might act normally to suppress synthesis of the transaminase. In that case, inhibitors of protein synthesis would be expected to prevent the increase in enzymic activity observed as a result of pretreatment with Dopa. In contrast to regulation of tyrosine transaminase in the liver [17] this aspect of regulation of the brain enzyme does not seem to involve the coenzyme, pyridoxal phosphate.

4.1.3 Adaptation involving coenzyme

This has been shown to occur in thiamine-deficient rats, and the adaptation is not restricted to the brain. Rats made deficient in thiamine have lowered mitochondrial pyruvate dehydrogenase activity in heart, liver and brain [18], even though the activity was measured in the presence of the coenzyme, thiamine pyrophosphate. Treatment of the animals with thiamine then results in a recovery of the enzymic activity to the normal value within 10 to 12 hr.; the recovery is prevented by prior treatment with actinomycin D or cycloheximide. Labelling of nucleic acids with uridine and of proteins with ^{14}C-leucine was stimulated by thiamine in the thiamine deficient animals (Table 4.2), where the response was greatest in the brain.

Table 4.2 Cerebral metabolic response to thiamine in thiamine-deficient rates [18]

Treatment	*Pyruvate dehydrogenase (% of control)*	*Uridine incorporation (% of control)*
Thiamine	160	2500
Thiamine + actinomycin D	106	100
Thiamine + cycloheximide	114	

4.1.4 Adaptation in response to hormones

Removal of the adrenals or the pituitary gland from adult rats is followed by an exponential decrease in glycerolphosphate dehydrogenase activity of the cerebral hemispheres and especially of the brain stem (Fig. 4.5). The activity depleted by adrenalectomy could be restored by injections of cortisol and that depleted after hypophysectomy was restored by injections either of cortisol or of adrenocorticotrophic hormone (ACTH). The enzymic activity in the livers of the same animals was not affected. Two types of glycerolphosphate dehydrogenase activity are present in the mammalian brain, as in other organs: a soluble cytoplasmic NAD^+-requiring enzyme (EC 1.1.1.8) and a mitochondrial flavoprotein enzyme (EC 1.1.95.5); only the cytoplasmic enzyme was found to be involved. The mitochondrial activity and other mitochondrial enzymes (malate dehydrogenase, isocitrate dehydrogenase) were not affected. Another cytoplasmic enzyme, lactate dehydrogenase, was also unchanged throughout. The results of these studies led the investigators to the conclusion of a specific regulation of cerebral glycerolphosphate dehydrogenase activity by corticosteroids [19].

The adult mammalian brain contains relatively little of this dehydrogenase activity and the maximum peak of activity is reached only late in the immature brain during development, at a stage coincident with later stages of myelination. In the developing rat brain this is between 5 and 6 weeks of age. The rate of increase in enzymic activity can be accelerated during the first two weeks of age by treatment with cortisol, and the normal development of the activity can be retarded by adrenalectomy or hypophysectomy (Fig. 4.5b) and also by X-ray irradiation. Myelination in the central nervous system is associated with glial cells (Chapter 2) and requires active lipid synthesis for which the glycerol 3-phosphate is needed. Cultured glial cells have accordingly been used to study this aspect of hormonal regulation: cortisol in the culture medium was then shown to induce glycerol-phosphate dehydrogenase activity, which was prevented by the presence of inhibitors of synthesis of RNA and proteins. The induction occurred with a variety of corticosteroids, but not with steroid sex hormones, insulin or cyclic AMP. Other enzymes tested were not affected by cortisol. During the course of these studies on glial cell cultures, lactate dehydrogenase activity (not affected by treatment with corticosteroids) was increased

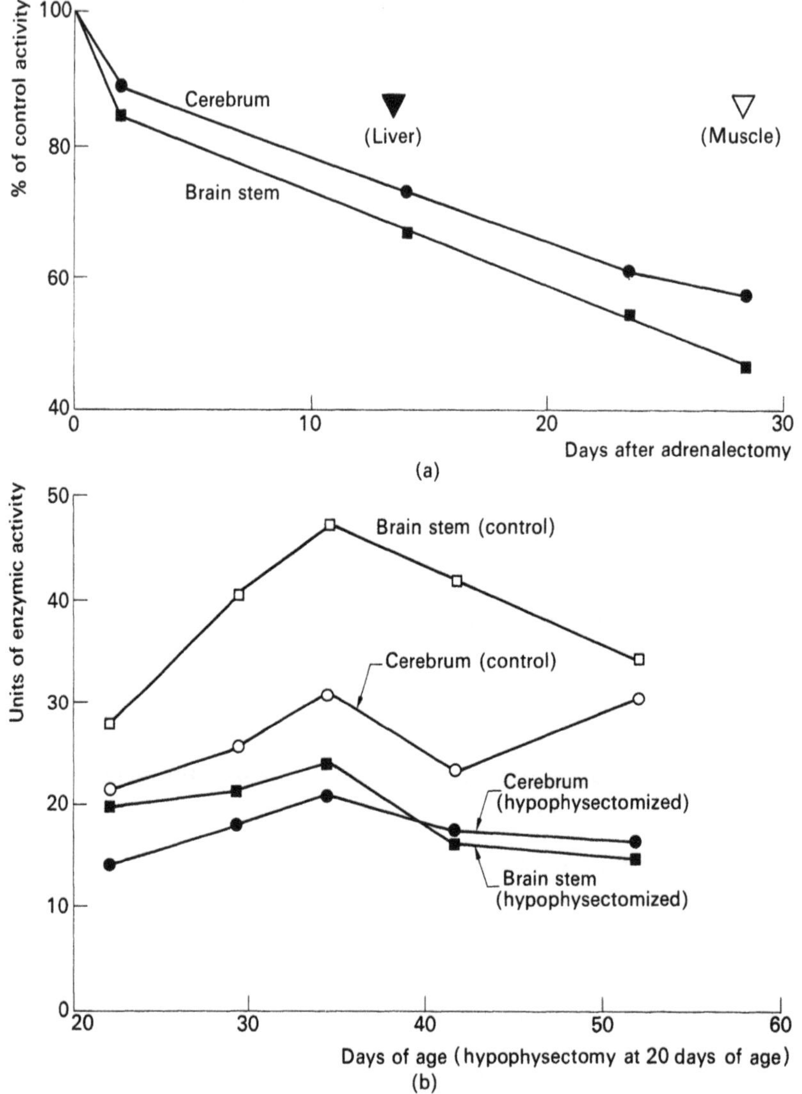

Fig. 4.5 Rat brain glycerolphosphate dehydrogenase activity [19].

significantly after the addition to the medium of adrenaline; the results of these studies have been interpreted in terms of regulation of glial lactate dehydrogenese activity by adrenaline.

The possible involvements of corticosteroids in cerebral adaptation are often assessed by following the effects of adrenalectomy or by treatment with dexamethasone or cortisone. Use of such techniques has indicated that these hormones contribute to regulation of many transmitter-synthesizing enzymes, such as tryptophan hydroxylase [20] and the phenylethanolamine *N*-methyltransferase which converts noradrenaline to adrenaline [21].

4.2 Adaptation to the environment

4.2.1 Light

The retina can be regarded as the external part of the nervous system and vision has been the sense most studied in terms of mechanisms of adaptation to the external environment. Fibres pass from the retinal ganglia, as the optic nerve, to the lateral geniculate body and thence to the visual cortex (Chapter 2).

Structural changes in the visual cortex are known to result from visual deprivation: if the animals are reared in darkness, morophological development of the visual cortex is impaired [22]. Light stimulation following visual deprivation also causes changes in the size and number of synapses in the visual cortex. Less clear have been the biochemical findings. Visual deprivation causes decreases in the content of RNA in retinal ganglion cells and in the visual cortex. The number of polyribosomes may also vary according to the environmental conditions. First exposure to light of rats reared in darkness results in a transient

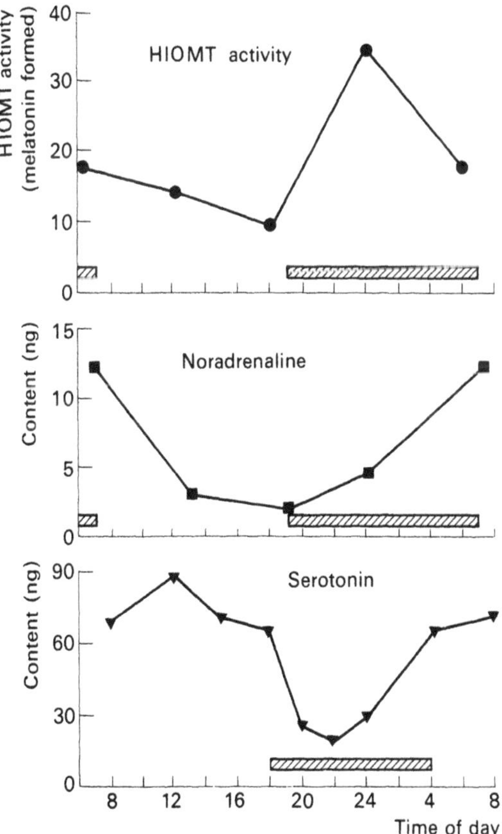

Fig. 4.6 Diurnal variation of serotonin, noradrenaline and 5-hydroxyindole O-methyl transferase activity (HIOMT) in rat pineal gland [26, 27]. The hatched bars give the periods of darkness.

increase in rates of ^{3}H-lysine incorporation into proteins of the visual cortex [22], and the brains of rats habituated to darkness have been reported to contain increased numbers of polysomes after exposure to light. In these animals, the light stimulation also caused increased rates of protein synthesis, which result correlates well with the increased number of polysomes [23]. However in a split-brain monkey preparation, unilateral visual stimulation had no effect on rates of protein synthesis in the subcellular fractions prepared from various regions [24]. Presumably the increase due to light stimulation does not occur in animals habituated to light, but requires prior visual deprivation. It should be noted that the proportion of ribosomes present as polysomes can also be changed by temperature variation and by convulsions [15].

Other biochemical observations in the visual cortex on adaptation to light include changes in assayable tubulin (Section 3.7.1) and in muscarinic receptors (Section 3.6.2) [25].

4.2.2 The pineal gland

The incidence of light causes an adaptive change in the biochemistry of the pineal gland, and provides an excellent example of enzyme induction caused by a sensory stimulus. The content of 5-hydroxytryptamine (serotonin) of the pineal gland was noted in Chapter 3 (Table 3.2) to be subject to diurnal variation, being high by day and low at night. In contrast, the serotonin of the hypothalamus remains at a relatively constant concentration and the variation in this amine seems peculiar to the pineal. Noradrenaline concentrations also vary in the reverse direction, being higher in the dark than in the light (Fig. 4.6). The mammalian pineal gland has evolved from a visual sensory organ of more primitive animals: in amphibian brains, for example, the equivalent is found as a photoreceptive area on the roof of the brain. The primitive pineal responded directly to light with direct transduction of the stimulus to nerve impulses. The mammalian pineal responds indirectly to light; the nerve impulse generated in the photoreceptors of the retina is carried by fibres of the optic nerves to the optic chiasma where the nerve tracts branch. One branch, the inferior accessory optic tract, carries the impulse to the pineal. Little is known of the nerve fibre connections from this tract to the pineal, but innervation of the gland is by sympathetic nerves from superior cervical ganglia [26]. The pineal appears to receive no direct input from the rest of the nervous system and does not send fibres to other parts of the brain.

The changes from light to darkness are associated with changes in another indole derivative, 5-methoxy-*N*-acetyltryptamine ('melatonin'). This is formed from serotonin by the acetylation and methyl transfer reactions shown in Fig. 4.7. The adaptive response to darkness involves release of serotonin and regulation of the enzyme which forms melatonin: 5-hydroxyindole *O*-methyl transferase (HIOMT; Fig. 4.7). The formation of serotonin itself is not susceptible in this way to changes in illumination; it is synthesised in the pineal at a relatively

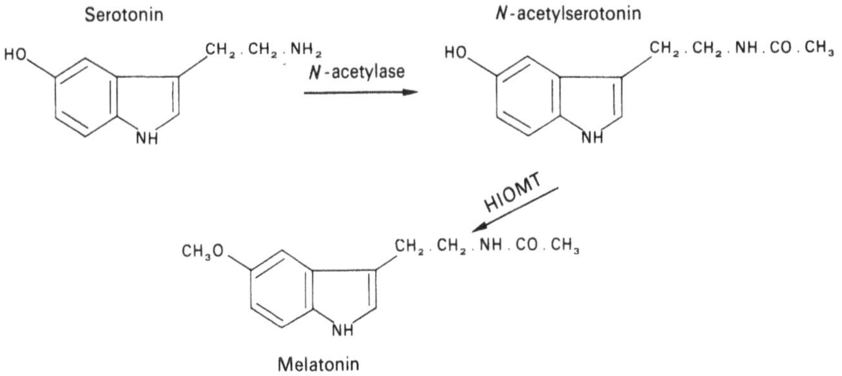

Serotonin

HO ... $CH_2 . CH_2 . NH_2$ (NH)

N-acetylase →

N-acetylserotonin

HO ... $CH_2 . CH_2 . NH . CO . CH_3$ (NH)

HIOMT ↙

CH_3O ... $CH_2 . CH_2 . NH . CO . CH_3$ (NH)

Melatonin

Fig. 4.7 Synthesis of melatonin (5-methoxy-*N*-acetyltryptamine). Serotonin is formed from tryptophan (Chapter 3). Melatonin is excreted after hydroxylation and conjugation in the liver as the sulphate glucuronide.

constant rate. However more serotonin is released from its storage sites in darkness than in light and so becomes more accessible to enzymes concerned with its further metabolism: HIOMT and monoamine oxidase. Part of the released serotonin is destroyed by the oxidase and part is converted to melatonin by HIOMT (Fig. 4.7). The methyl donor for methylation of N-acetylserotonin is S-adenosylmethionine. As noted above, the level of noradrenaline is out of phase with the pineal content of serotonin. During daylight hours, noradrenaline is low and serotonin is high. HIOMT activity is also lower during the day which indicates that the illumination indirectly causes inhibition of the enzymic activity. It is not clear which of two alternative processes occurs: either the arrival of the nerve impulse results in inhibition of release of a transmitter which normally stimulates or induces HIOMT, or it results in increased release of a transmitter which inhibits HIOMT. Thus it is not certain if the increase in HIOMT can properly be regarded as enzyme induction because the change in activity could equally be due to suppression of enzyme synthesis in daylight. Some indications that induction may occur have come from studies on the effects of noradrenaline in increasing melatonin synthesis in pineal glands maintained in an organ bath. The effects were blocked by cycloheximide. The serotonin *N*-acetyl transferase of Fig. 4.7 also responds to the dark (it is 30–50 times more active in the dark than in daylight) and to catecholamines. The activity of the enzyme in the pineal gland *in vivo* is activated by catecholamines (Dopa, noradrenaline and adrenaline) and by amine-oxidase inhibitors which prevent breakdown of catecholamines. Involvement of cyclic AMP was indicated by the observation of activation of the *N*-acetyl transferase by theophylline (which inhibits breakdown of cyclic AMP by phosphodiesterase). The increased enzymic activity was prevented by treatment with inhibitors of protein synthesis and by propranolol (which blocks adrenergic β-receptors; Chapter 3). The results of these studies suggest that new transferase enzyme is synthesised as a consequence of adrenergic stimulation of pineal β-receptors, and involves activation of formation of cyclic AMP [28]. Thus the two enzymes involved in melatonin formation from serotonin (Fig. 4.7) are known to respond to catecholamines: it seems reasonable to conclude from the evidence available at present, that the light-stimulated nerve impulse releases noradrenaline from the pineal, thus making less available for enzyme induction [29].

The role of melatonin in mammals seems to be essentially inhibitory: it inhibits thyroid function and sexual maturation in both males and females. It also appears to suppress the secretion of luteinizing hormore (LH) and it is possibly involved in regulation of adrenocortical function although the results of research on this aspect are confusing and sometimes contradictory [26].

4.3 Drug tolerance and dependence

The phenomenon of drug addiction involves a process whereby a chemical substance which is foreign to the body becomes an intrinsic part, essential for continued maintenance of normal function. Normally, first acquaintance with an addictive drug elicits a positive response by the body. This may be an unpleasant or toxic response or, as in the case with analgesics and certain stimulants, the response may be pleasurable or euphoric. If, on continued exposure to the drug, these responses are lessened or disappear, tolerance to the drug has developed. If the individual is only 'well' during continued exposure to the drug, and becomes 'ill' when it is withdrawn, then a state of physical dependence on the drug has been set up.

Tolerance is usually assessed by measuring the quantitative response to standard doses of the drug. The response of a tolerant animal is expressed as a percentage of its response before exposure to the drug or of the response of a control group of animals. Physical dependence is assessed by measuring the degree of illness experienced on withdrawal of the drug or on administration of a specific antagonist to the drug.

The mechanism of the body's response (tolerance or dependence) to a foreign drug may be similar in principle to the mechanisms which underly the effects of many of the body's normal endogenous chemicals in modifying physiological and biochemical function by affecting the rate of synthesis of a specific enzyme; the chemical may bear no obvious structural resemblance to the chemicals (substrates and products) normally involved with that enzyme. Many workers have therefore approached the phenomenon of drug tolerance and dependence on the assumption that induction or repression of enzyme synthesis is involved [30]. A molecule which acts as an enzyme inducer needs to have no structural relationship with the enzyme because it presumably acts at sites, remote from the enzyme itself, to dissociate specific repressor protein molecules from their sites on the cellular DNA in the nucleus. It seems possible also that, just as enzyme synthesis is modifiable in this way, the synthesis of receptor proteins could also be affected, so as to render the receptor either more or less sensitive to its transmitter, and we need therefore to think in terms of induction of synthesis of non-enzyme proteins as well as of enzymes. The major criterion for assessing *de novo* synthesis of proteins and enzymes rests on the use of inhibitors of protein synthesis; while these may be specific in their sites of interaction within the sequence of RNA and protein synthesis, they are quite unspecific in that they are likely to inhibit protein synthesis indiscriminately. A table of some of the side effects of these inhibitors is given by Shuster [30]. One means of circumventing the grosser artefacts which might result from this lack of specificity is to use methods of developing tolerance and dependence in animal models sufficiently rapidly (within a few hours) so that tests with the inhibitors of protein synthesis are less likely to have indiscriminate effects. This approach has been used successfully by Cox and his co-workers studying the morphine tolerance and dependence [31] which follows.

4.3.1 Morphine

Addiction to morphine, the alkaloid principle prepared from opium, often results from its legitimate prescribed use as an analgesic and a great amount of effort has been expended in attempts to find a pharmacological agent which would prevent the addiction without impairing analgesic function. Great hope that this may be soon achieved arose from the discovery of the endogenous opiates, enkephalins and endorphins (see Chapter 3). However, though they react with central morphine receptors and produce analgesia, the evidence currently available indicates that they too are addictive. While it may prove feasible to produce analogues which are analgesic and non-addictive, this may depend on whether the analgesia is mediated by the same mechanisms as are involved in the euphoria which seems an integral part of the addiction; only if these are different are our hopes of separating the responses likely to be realised. It has proved relatively easy to produce animal models: the rat can be made tolerant to morphine, either by subcutaneous implantation of the drug as a pellet, or by intravenous infusion, where tolerance can be shown to be developed within a few hours (Fig. 4.8). The animals are then insensitive to the analgesic effect of some 20 times the normal dose. Their sensitivity to pain in these studies is usually measured by their response to a tail-flick. The development of tolerance to the drug is not due to increased destruction or excretion of the drug, and can be prevented by inhibitors of protein synthesis. The results, Fig. 4.8b, show that actinomycin D does not overcome tolerance already established; it only affects the rats during the period when tolerance is developing. Use of a range of inhibitors of RNA and protein synthesis showed the development of tolerance to depend on synthesis of new messenger RNA [31]. This knowledge has led to attempts to identify the enzymes (or proteins) whose synthesis might be induced or suppressed by morphine. The brains of tolerant animals show changes in

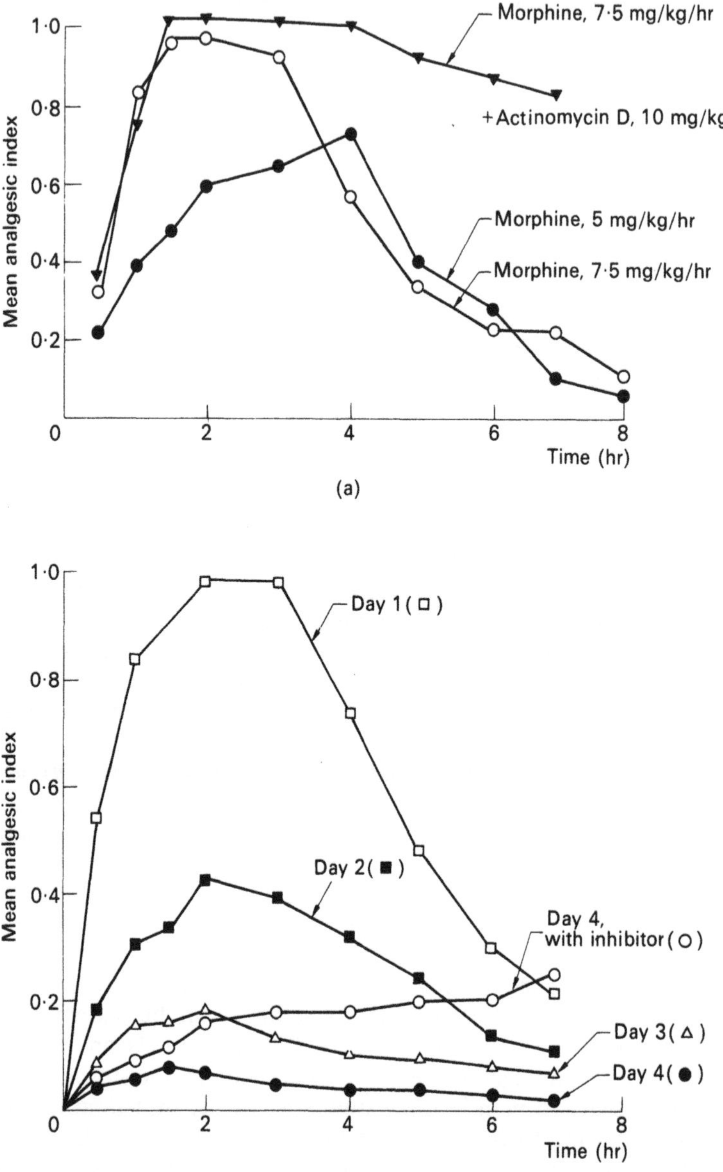

Fig. 4.8 Development of morphine tolerance in rats [31]. **(a):** Morphine was infused intravenously into rats and the development of tolerance (when the analgesic effects of the drug were lessened) was prevented by the presence of the inhibitor, actinomycin D. **(b):** The tolerance to morphine increases with subsequent days of infusion. The presence of actinomycin D in the infusion fluid on the 4th day of infusion (○) shows that it affects the tolerance only during the period when it is developing, but does not overcome the tolerance previously established.

respiration and in the turnover of phospholipids, especially of triphosphoinositide, a phospholipid component of synaptic membranes. Various neurotransmitters, implicated by different workers, include acetylcholine, serotonin, noradrenaline and GABA, from observations that their antagonists affect the rate of onset and the degree of tolerance attained. The development of physical dependence to the drug, based on physical and behavioural manifestations which occur when the drug is withdrawn, has also been suggested to involve one or all of these transmitters. The one enzyme found to be increased during the development of tolerance so far is tryptophan hydroxylase (involved in formation of serotonin, Chapter 3). Cyclic AMP seems to be involved in the effects of serotonin and its antagonists on the development of tolerance to and dependence on morphine [32].

Interactions of all neurotransmitters might be affected: results of a study on the Ca^{2+}-activated ATPase suggest this could be the case. Ca^{2+} is an essential requirement for the release of all neurotransmitters (Chapter 3): the synaptosomal Ca^{2+}-ATPase prepared from normal rats was inhibited by morphine *in vitro* in contrast to the enzyme from morphine-tolerant rats, which remained unaffected by exogenous morphine [33]. If the response to morphine is on transmitter release, then the specificity of the transmitter involved may be related to the more sensitive regions of the brain. Lack of interaction of transmitter with receptor could, from the mechanisms discussed in previous sections of this chapter, result in increased synthesis of specific enzymes, such as the increase in tryptophan hydroxylase activity noted above. This is not the only likely mechanism by which morphine might act: blockage of transmitter receptors could also result in changes in synthesis of enzymes involved in transmitter metabolism. The use of drugs which interfere with serotonergic transmission may not distinguish between action on transmitter release and action on receptors. *p*-Chlorophenylalanine (PCPA), which inhibits serotonin synthesis, 5, 6-dihydroxytryptamine (DHT), which destroys serotonergic nerve endings, and methergoline, which blocks serotonin receptors, all inhibit the development of tolerance and of dependence [32].

The role of the endogenous opiates in pain perception and analgesia is essentially still a matter for speculation, as is the possibility that their function is related to acupuncture. Current research on the mechanisms of function of the enkephalins and endorphins will determine whether they behave as transmitters or modulators, and how this is related to release of other transmitters. Such research should result in further knowledge on the cerebral responses to morphine, in terms of receptor sensitivity or enzyme induction, and provide more insight into the mechanisms involved in addiction.

4.3.2 Amphetamines

The group of amphetamine stimulants have long been recognised as leading to psychological dependence, but unequivocal evidence for the development of physical dependence is not so well established. The amphetamines are known to inhibit mono-amine oxidase and to interfere with re-uptake of catecholamines (Chapter 3). Prolonged abuse can lead to the development of psychotic disturbances which tend to disappear on withdrawal of the drug. Severe addiction, with a high relapse rate, was noted by Connell [34] to be similar in some aspects to alcohol addiction. Amphetamine psychosis is a relatively rare consequence of abuse but in Japan, encouraged use of amphetamines to increase industrial productivity after the end of the 1939–1945 war led to dependence and psychosis of almost epidemic proportions. Animal models for psychological dependence, monitored by self-administration of the drug, proved relatively easy to develop; models of physical dependence have not proved so easy to assess, due to variations in behaviour on withdrawal. The major emphasis on the possible biochemical interactions of the amphetamines has been on adrenergic transmission [15]. Hyperactivity and stereotyped behaviour in experimental animals seems to be associated with noradrenaline and dopamine: the effects are antagonised by propranolol (Chapter 3) and by e.g. α-methyltyrosine, which blocks catecholamine formation by inhibiting tyrosine hydroxylase. Induction of tyrosine hydroxylase by catecholamines (above) also occurs on administration of amphetamines, and the increase in enzyme activity is prevented by pretreatment with cycloheximide [35]. Such effects are associated with relatively short-term administration of the drug. Long-lasting effects on mono-amine oxidases were observed when methamphetamine was given to guinea pigs for 3 to 10 weeks: the 30% decrease in activity was reversed on withdrawal of the drug, when it reached a level some 30% above normal. Acute administration of the drug resulted in increased cerebral respiration, which was decreased on chronic administration [36]. The effects of amphetamine on cerebral oxidative metabolism were assumed to be secondary to the effects on catecholamines, which were expected to cause changes in mobilisation of glycogen by phosphorylase activation mediated by cyclic AMP. This does not seem to be confirmed by recent studies. Though catecholamine activation, mediated by cyclic AMP, of cerebral glycogen phosphorylase is known to occur, the increased rates of cerebral glycolysis after acute treatment with amphetamines cannot be explained solely in terms of increased glycogen breakdown. Some evidence for dependence emerged from these studies. The increase in glycolytic rates shown after acute administration disappeared on chronic treatment with the drug, when the rates reverted to normal. Withdrawal of the drug from chronically-treated rats resulted in rates of glycolysis significantly lower than normal, which were similar to those seen in animals treated with depressant drugs, such as the barbiturates (Fig. 4.9). Chronic administration of amphetamine to rats also causes profound changes in cerebral tyrosine hydroxylase activity [38, 39], only if the rats fail to develop tolerance to the anorexic effects of the drug. The activity recovered within 1 day after withdrawal of the drug (Fig. 4.10).

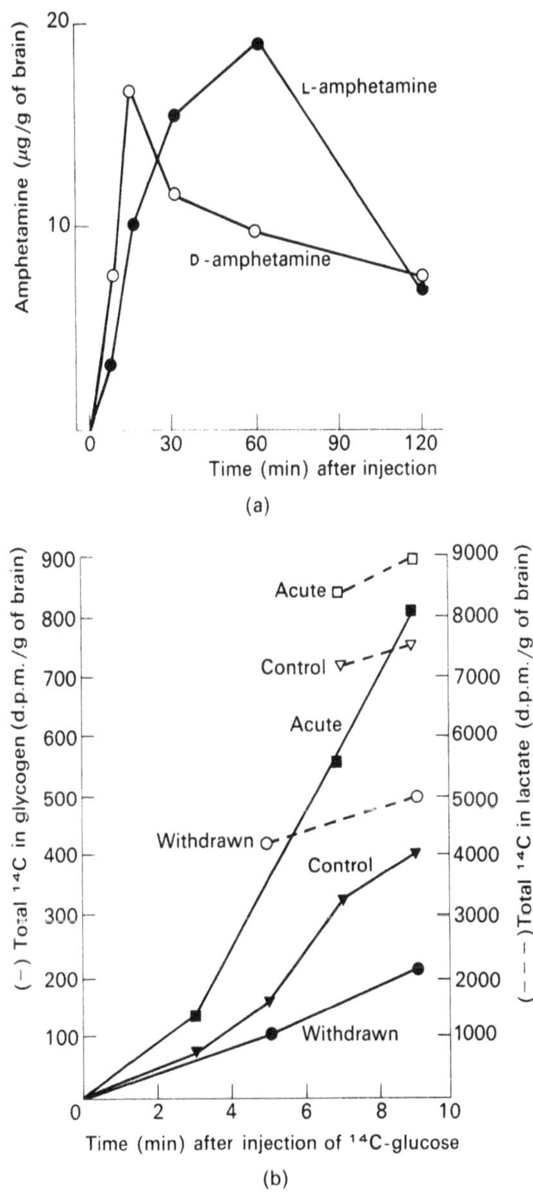

Fig. 4.9 Some effects of amphetamines. **(a):** Concentrations of D- or L-amphetamine in the brain after a single intraperitoneal injection, of 15 mg./kg. body weight, into mice [40]. **(b):** Incorporation of ^{14}C from glucose into glycogen (—) and lactate (---) in the brains of rats treated with methamphetamine. 'Acute': one intraperitoneal injection (5 mg./kg.) 1 hr. beforehand; 'chronic': the drug was administered in the drinking water in increasing concentrations over a 3 week period. Initial intake was 5 mg./kg./day and final intake was 40 mg./kg./day; 'withdrawn': as for chronic administration but the drug was absent from the drinking water for the final 24 h. period. The chronic groups gave results identical to those of the control groups of rats [37].

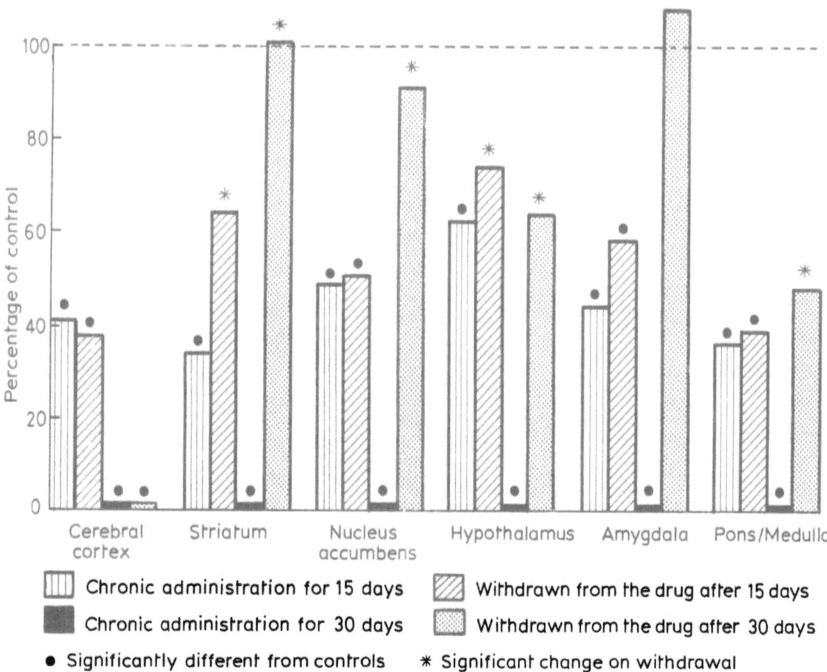

Fig. 4.10 Tyrosine hydroxylase activity expressed as percentage of activity in sham-treated controls, in various regions of rat brain after treatment with methamphetamine [39].

The activity was unchanged in rats which developed tolerance. Contents of noradrenaline and dopamine, and of their metabolites, were also decreased in conditions where tyrosine hydroxylase activity was decreased. Thus biochemical interactions of the amphetamines have been shown to involve synthesis, transport and destruction of catecholamines and also to affect glycolysis in the brain. While the relationship between the two areas of metabolism remains obscure, both can be used as criteria of the development of dependence, with the striking metabolic changes which occur on withdrawal (Figs. 4.9, 4.10).

4.3.3 Ethanol

Adaptation to ethanol involves both increased peripheral metabolism of the drug and alterations in cerebral function. The major metabolites which appear in the body fluids as a result of oxidation of ethanol, predominantly in the liver, are acetaldehyde and acetoacetate. The development of physical dependence on alcohol in man is shown in the hallucinations and tremors which occur on its withdrawal. The most pronounced enzymic adaptation to alcohol which has been observed is the increase in liver alcohol dehydrogenase, an NAD^+-requiring enzyme which produces acetaldehyde. The increase is thought to be due to substrate induction of enzyme synthesis, because it can be prevented in rats by cycloheximide or by actinomycin D [15].

While the toxic effects of ethanol can be attributed largely to the unpleasant effects of acetaldehyde, the hallucinations and other physical and pharmacological effects of dependence, and particularly of withdrawal, suggest specific neural interactions. In particular, the hallucinations have stimulated an interest in the metabolism of biogenic amines; increased urinary excretion of tryptamine pointed to a possible effect on serotonin metabolism [41]. While increased dopamine breakdown occurs in the striatum with acute ethanol treatment, this disappears on chronic administration, when changes in sensitivity of dopamine receptors have been observed. Also chronic treatment resulted in increased levels of metenkephalin there [42]. Despite many studies on the effects of ethanol on the storage, release and oxidation of serotonin and catecholamines, their involvement in the adaptive response by the brain to ethanol remains unclear [43, 44]. It seems appropriate to note that oxidation of alcohol in the brain by alcohol dehydrogenase, an NAD-linked enzyme, will tend to lower the ratio of NADH to NAD^+ which in turn would affect a wide variety of intermediary metabolic processes [45].

4.4 Learning and memory as adaptive processes?

The recent increase in our understanding of the genetic code and transference of biological information has prompted renewed interest in possible macromolecular mechanisms of learning and memory. Memory fixation has been observed to occur in 2 phases: 'short-term' memory which lasts about 30 min., which is then fixed as 'long-term' memory. Studies based on lesions of various brain regions and consequent disturbance of memory have indicated areas of the hippocampus and amygdala of the limbic system to be associated with short-term memory, and the 'association regions' of the cortex (Chapter 2) with long-term memory. Due to the knowledge that transfer of genetic information involves nucleic acid and protein molecules, much emphasis has been placed by biochemists on these. The cautions necessary in interpreting the results from the use of inhibitors of protein synthesis (Section 4.3) should be kept in mind but it seems to be agreed that inhibitors such as puromycin and cycloheximide, while having no effect on short term memory, can prevent fixation as long-term memory [46].

Most current theories on memory fixation involve aspects of synaptic modification, i.e. an adaptive response of synaptic connections to a learned situation either by activating hitherto dormant inactive synapses or by rendering previously active synapses inactive. This could be by modifications in proteins (or lipids) of the lipoprotein synaptic membranes so as to change their conformation and permeability properties, or to change the sensitivity of specific receptor sites. Such thoughts have led to searches for 'memory molecules' which might cause such changes, so far usually as proteins or peptides. Many experiments which are based on changes in the random incorporation of isotopically labelled amino acids into large and diffuse protein fractions of the brain seem difficult to accept: it appears unlikely that a change in rate of synthesis of a proteins or small family of proteins, in response to the learning of a relatively simple task, could be easily detected since all of the proteins in the tissue sample will incorporate the label. Any change in a small proportion of the proteins (or lipids or nucleic acids, as the case may be) seems almost certain to be masked by the lack of change in the majority of the related macromolecules. More acceptable is the attempt to study the labelling of specific proteins or other molecules peculiar to the nervous system; some progress has been made in this direction [47].

One major handicap in deciding which type of experiment is likely to prove fruitful is the uncertainty of the 'memory sites'. While it is accepted that certain areas noted above are associated with memory, it is not at all clear if the sites are specific. Animal lesioning experiments indicate little or no cellular specificity whereas the evidence from studies on man suggests the opposite [48]. It has been difficult to demonstrate association of any specified localised part of the brain with any particular learning or memory situation, so the possibility of an analogy to the hologram has been raised. This is at least compatible with the idea of the relatively wide distribution of a memory molecule, which reacts on numerous synapses, rather than activation of specific synapses. This is discussed in detail by Rose [48] who proposes a combined hypothesis — the 'redundant network, modifiable synapse' theory – which includes the ability of the brain to replace loss of function in one site by the same function elsewhere, thus allowing the whole of the organ to continue to function if one part should fail, with the basic assumption of synaptic modification noted above.

References

[1] Kandel, E. R. and Spencer, W. A. (1968), 'Cellular neurophysiological approaches in the study of learning'. *Physiol. Revs.* **48**, 66–134.

[2] Thoenen, H. (1972), 'Neuronally mediated enzyme induction in adrenergic neurones and adrenal chromaffin cells'. *Biochem. Soc. Symp.* **36**, 3–15.

[3a] Mueller, R. A., Thoenen, H. and Axelrod, J. (1969), 'Increase in tyrosine hydroxylase activity after reserpine administration'. *J. Pharm. exp. Ther.* **169**, 74–79.

[3b] Thoenen, H. Meuller, R. A. and Axelrod, J. (1969), 'Trans-synaptic induction of adrenal tyrosine hydroxylase'. *J. Pharm. exp. Ther.* **169**, 249–254.

[4] Black, I. B., Hendry, I. A. and Iversen, L. L. (1971), 'Trans-synaptic regulation of growth and development of adrenergic neurones in a mouse sympathetic ganglion'. *Brain Res.* **34**, 229–240.

[5] Mackay, A. V. P. and Iversen, L. L. (1972), 'Increased tyrosine hydroxylase activity of sympathetic ganglia cultured in the presence of dibutyryl cyclic AMP. *Brain Res.* **48**, 424–426. 'Trans-synaptic regulation of tyrosine hydroxylase activity in adrenergic neurones: effect of potassium concentration on cultured sympathetic ganglia'. *Naunyn-Schmiedeberg's Arch. Pharmacol.* **272**, 225–229.

[6] Thoenen, H. (1974, 1975), 'Trans-synaptic enzyme induction', *Life Sci.*, **14**, 223–235; *Adv. Neurol.*, **6**, 67–71.

[7] Klee, G. B. and Sokoloff, L. (1967), 'Changes in the D(-)-β-hydroxybutyrate dehydrogenase activity during maturation in the rat'. *J. Biol. Chem.* **242**, 3880–3883.

[8a] Pull, I. and McIlwain, H. (1971), '3-Hydroxy-butyrate dehydrogenase of rat brain on dietary change and during maturation'. *J. Neurochem.* **18**, 1163–1165.

[8b] Williamson, D. H., Bates, M. W., Page, M. A. and Krebs, H. A. (1971), 'Activities of enzymes involved in acetoacetate utilisation in adult mammalian tissues.' *Biochem. J.* **121**, 41–47.

[9] Hawkins, R. A. (1971), 'Uptake of ketone bodies by rat brain *in vivo*,' *Biochem. J.* **121**, 17P.

[10] Owen, O. E., Morgan, A. O., Kemp. H. G., Sullivan, J. M., Herrera, M. G. and Cahill, G. F. (1967), 'Brain metabolism during fasting'. *J. Clin. invest.* **46**, 1589–1595.

[11] Itoh, T. and Quastel, J. H. (1970), 'Acetoacetate metabolism in infant and adult rat brain *in vitro*'. *Biochem. J.* **116**, 641–655.

[12] Gjedde, A. and Crone, C. (1975), 'Induction processes in blood-brain transfer of ketone bodies during starvation', *Am. J. Physiol.*, **229**, 1165–1169.

[13] Kraus, P. (1970), 'Substrate induction of mouse brain L-glutamate decarboxylase'. *Hoppe-Seyler's Z. Physiol. Chem.* **349**, 1425–1427.

[14] Sze, P. Y. (1970), 'Possible repression of L-glutamic acid decarboxylase by gamma aminobutyric acid in developing mouse brain'. *Brain Res.* **19**, 322–325.

[15] McIlwain, H. and Bachelard, H. S. (1971), *Biochemistry and the Central Nervous System.* 4th. ed., London: Churchill.

[16] Gibb, J. W. and Webb, J. G. (1969), 'The effects of reserpine, α-methyltyrosine and L-3, 4-dihydroxyphenylalanine on brain tyrosine transaminase'. *Proc. Nat. Acad. Sci. U.S.* **63**, 364–369.

[17] Black, I. B. and Axelrod, J. (1968), 'Elevation and depression of hepatic tyrosine transaminase activity by depletion and repletion of norepinephrine'. *Proc. Nat. Acad. Sci. U.S.*, **59**, 1231–1234.

[18] Reinauer, H. and Hollmann, S. (1969), 'The co-enzyme-dependent induction of pyruvate dehydrogenase in thiamine deficiency', *Hoppe-Seyler's Z. physiol. Chem.* **350**, 40–50.

[19] De Vellis, J. and Inglish, D. (1968), 'Hormonal control of glycerol phosphate dehydrogenase in rat brain'. *J. Neurochem.* **15**, 1061–1070; Abstr. Second Intern. Neurochem. Meeting (1969), Milan: Tamburini, p. 152.

[20] Rastogi, R. B. and Singhal, R. L. (1978), 'Adrenocorticoids control 5-hydroxytryptamine metabolism in rat brain', *J. Neural Transm.*, **42**, 63–71.

[21] Turner, B. B., Katz, R. J. and Carroll, B. J. (1979), 'Neonatal corticosteroid permanently alters brain activity of epinephrine-synthesizing enzyme in stressed rats', *Brain Res.*, **166**, 426–430.

[22] Rose, S. P. R. (1968), 'Biochemical aspects of memory mechanisms' in *Applied Neurochemistry* (ed. Davison, A. N. and Dobbing, J.), Blackwell, Oxford, pp. 356–376.

[23] Appel, S. H., Davis, W. and Scott, S. (1967), 'Brain polysomes: response to environmental stimulation'. *Science*, **157**, 836–838.

[24] Metzger, H. P., Cuenod, M., Grynbaum, A and Waelsch, H. (1967), 'The effect of unilateral visual stimulation on synthesis of cortical proteins in each hemisphere of the split-brain monkey'. *J. Neurochem.* **14**, 183–187.

[25] Rose, S. P. R. (1978), 'Macromolecular mechanisms and long-term changes in behaviour', *Biochem. Soc. Trans.*, **6**, 844–848.

[26] Wurtman, R. J., Axelrod, J. and Kelly, D. E. (1968), *The Pineal*, New York, A.P.

[27] Axelrod, J., Shein, H. M. and Wurtman, R. J. (1969), 'Stimulation of ^{14}C-melatonin synthesis from ^{14}C-tryptophan by noradrenaline in rat pineal organ culture'. *Proc. Nat. Acad. Sci. U.S.* **62**, 544–549.

[28] Deguchi, T. and Axelrod, J. (1972), 'Induction and superinduction of serotonin N-acetyl transferase by adrenergic drugs and denervation in rat pineal organ'. *Proc. Nat. Acad. Sci. U.S.* **69**, 2208–2211.

[29] Wolstenholme, G. E. W. and Knight, J. (ed.), (1971), *The Pineal Gland.* London: Churchill.

[30] Shuster, L. (1971), 'Tolerance and physical dependence' in *Narcotic Drugs* (ed. Clouet, D. H.), New York: Plenum, pp. 408–423.

[31a] Cox, B. M., Ginsburg, M. and Osman, O. H. (1968), 'Acute tolerance to narcotic analgesis drugs in rats'. *Br. J. Pharmac. Chemother.* **33**, 245–256.

[31b] Cox, B. M. and Osman, O. H. (1970), 'Inhibition of the development of tolerance to morphine in rats by drugs which inhibit ribonucleic acid or protein synthesis'. *Br. J. Pharmac.* **38**, 157–170.

[32] Way, E. L., Ho, I. K. and Loh, H. H. (1973), 'Some biochemical aspects of morphine tolerance and physical dependence'. *Internat. Neurochem. Meeting, Tokyo*, **3**, 32–33, 478.

[33] Kaneto, H., Koku, T. and Koida, M. (1973), 'Inhibitory effect of morphine on synaptosomal Ca^{2+}-activated ATPase and its absence in morphinized mice'. *Internat. Neurochem Meeting, Tokyo*, **3**, 479.

[34] Connell, P. H. (1958), *Amphetamine Psychosis.* London, Chapman and Hall.

[35] Mandell, A. J. and Morgan, M. (1970), 'Amphetamine induced increase in tyrosine hydroxylase activity'. *Nature*, **227**, 75–76.

[36] Utena, H., Ezoe, T., Kato, N. and Hada, H. (1959), 'Effects of chronic administration of methamphetamine in enzymic patterns in brain tissue'. *J. Neurochem.* **4**, 161–169.

[37] Manning, D. H., Strang, R. H. C. and Bachelard, H. S. (1974), 'Changes in cerebral carbohydrate metabolism in the rat after acute and chronic treatment with, and withdrawal of, methampetamine'. *Biochem. Pharmacol.* **23**, 1205–1209.

[38] Javoy, F., Agid, Y., Bouvet, D. and Glowinski, J. (1974), '*In vivo* estimation of tyrosine hydroxylation in the dopaminergic terminals of the rat neostriatum', *J. Pharm. Pharmac.*, **26**, 179–185.

[39] Bardsley, M. E. and Bachelard, H. S. (1981), 'Catecholamine levels and tyrosine hydroxylase activity in rat brain regions after chronic treatment with, and withdrawal of, methampethamine'. *Biochem. Pharmac.*, in press.

[40] Benakis, A. and Thomasset, M. (1969), 'Metabolism of amphetamines and their interaction with other drugs' in *Abuse of central stimulants* (ed. Sjöqvist, A. and Tottie, M.). Almqvist and Wiksell, Stockholm, pp. 409–434.

[41] Schenker, V. J., Kissin, B., Maynard, L. S. and Schenker, A. C. (1967), 'The effects of ethanol on amine metabolism in alcoholism' in *Bio-*

chemical factors in alcoholism (ed. Maickel, R. P.). Pergamon, London, pp. 39–52.

[42] Reggiani, A., Barbaccia, M. L., Spano, P. F. and Trabucchi, M. (1980), 'Dopamine metabolism and receptor function after acute and chronic ethanol', *J. Neurochem*, **35**, 34–37.

[43] Ritzmann, R. F. and Tabakoff, B. (1976), 'Ethanol, serotonin metabolism and body temperature', *Ann. N. Y. Acad. Sci.*, **273**, 247–255.

[44] Blum, K. (ed.) (1977), *Alcohol and Opiates*, Academic Press, New York.

[45] Krebs, H. A. (1980), 'The effects of alcohol on metabolic processes', In: *Addiction and Brain Damage* (ed. Richter, D.), pp. 11–16, Croom Helm, London.

[46] Agranoff, B. (1973), in *Macromolecules and Behaviour* (ed. Ansell, G. B. and Bradley, P.), Macmillan, London.

[47] Hyden, H. (1973), in *Macromolecules and Behaviour* (ed. Ansell, G. B. and Bradley, P.), Macmillian, London.

[48] Rose, S. P. R. (1973), *The conscious brain*. Weidenfeld and Nicolson, London; Rose, S. P. R. and Longstaff, A. (1981), 'Neurochemical aspects in learning and memory', In: *Neurobiology of Learning and Memory* (eds. McGough, J. and Thompson, R.), Plenum, New York.

Index

GPSR Compliance
The European Union's (EU) General Product Safety Regulation (GPSR) is a set of rules that requires consumer products to be safe and our obligations to ensure this.

If you have any concerns about our products, you can contact us on

ProductSafety@springernature.com

In case Publisher is established outside the EU, the EU authorized representative is:

Springer Nature Customer Service Center GmbH
Europaplatz 3
69115 Heidelberg, Germany

www.ingramcontent.com/pod-product-compliance
Ingram Content Group UK Ltd.
Pitfield, Milton Keynes, MK11 3LW, UK
UKHW041822200726
13854UKWH00001BA/441
9789400959422